HOME PLUMBING HANDBOOK

by Charles N. McConnell

MACMILLAN PUBLISHING COMPANY
NEW YORK

Collier Macmillan Publishers
LONDON

Third Edition

Macmillan Publishing Company
866 Third Avenue, New York, NY 10022
Collier Macmillan Canada, Inc.

Library of Congress Cataloging-in-Publication Data
McConnell, Charles N.,
    Home plumbing handbook.
    Includes index.
    1. Plumbing—Amateurs' manuals. I. Title.
TH6124.M3   1984   696'.1   84-12506
ISBN 0-672-234130

Macmillan Books are available at special discounts for bulk
purchases for sales promotions, premiums, fund-raising, or
educational use. For details, contact:

        Special Sales Director
        Macmillan Publishing Company
        866 Third Avenue
        New York, NY 10022

10 9 8 7 6 5

Printed in the United States of America

Preface

Do-it-yourself is like learning a trade, or rather like learning several trades. The more you do, the more you learn, providing you have good instructions. Retailers, recognizing the demand for do-it-yourself materials, now devote entire stores to this market. All types of building materials used in the home are available, and usually at discount prices.

Now the knowledge of how-to-do-it is also available. Clear and concise *detailed* instructions for repairing, installing, or replacing the plumbing piping and fixtures in every home as well as instructions for preventive maintenance are found in *Home Plumbing Handbook*.

The uses and installation methods for PVC and CPVC piping and fittings, which require no special tools for installation, are explained. PVC and CPVC piping and fittings are used extensively for home repairs and modernization projects as well as in new home construction.

Repair and replacement of faucets, toilet tank fittings, plumbing fixtures, and copper, steel, and cast-iron pipe are only a few of the subjects covered in this book. And they are all explained *in detail*, with illustrations to cover any questions which may arise.

Home Plumbing Handbook is an invaluable reference book for every homeowner.

Contents

Acknowledgments

The author wishes to thank the following companies for their assistance in furnishing information and drawings on their products.

American Standard
P.O. Box 2003
New Brunswick, NJ 08903

Colonial Engineering Inc.
4000 Metzger Road
Ft. Pierce, FL 33450

Delta Faucet Co.
P.O. Box 40980
Indianapolis, IN 46280

Flint & Walling Co.
Kendallville, IN 46755

Fluidmaster, Inc.
P.O. Box 4204
Anaheim, CA 92803

IN-SINK-ERATOR
Emerson Electric Co.
4700 21st Street
Racine, WI 53406

Jet Aeration Co.
750 Alpha Drive
Cleveland, OH 44143

J. H. Industries
1712 Newport Circle
Santa Ana, CA 92705

Kohler Co.
Kohler, WI 53044

Nibco
Elkhart, IN 46514

Radiator Specialty Co.
P.O. Box 10628
Charlotte, NC 28237

Ridge Tool Co.
400 Clark Street
Elyria, OH 44032

Stanadyne Moen Division
377 Woodland Avenue
Elyria, OH 44036

The author also wishes to thank his wife, Joyce, for her help in preparing the manuscript and proofreading the text.

Some of the subjects covered in the book require disconnection and reconnection of electrical wiring or the installation of new wiring. Before starting on any project where electrical wiring is involved, *remove* the fuse or trip the circuit breaker in the electrical circuit. The installations and repairs explained in this book must conform to local codes and laws.

About the Author

Charles N. McConnell is a licensed Master Plumber who was actively engaged in the plumbing trade before he retired and started writing do-it-yourself books. He is the recipient of an award for over forty years' continuous membership in the United Association of Journeymen Plumbers and Steamfitters. He has designed, supervised, and installed plumbing, heating, and air-conditioning installations in residential, commercial, industrial, and public buildings and has also trained many apprentices to the plumbing trade. He is the author of the three-volume *Plumbers and Pipefitters Library* (Bobbs-Merrill) and *Building an Addition to Your Home* (Prentice-Hall).

CHAPTER 1

General Plumbing Information

The plumbing systems which we in the United States have in our homes are the most modern in the world. They are so trouble-free that we tend to take our plumbing for granted. Take the time now to learn some basic facts about the plumbing system in your home. When repairs become necessary or if a remodeling project requires some plumbing changes, you will be able to cope with the problems.

Technically, the plumbing system of a building consists of the water supply to the fixtures and the drainage piping from the fixtures to the building sewer. Gas or oil piping, while often used, is not actually part of the plumbing system, but of the piping system. The plumbing and piping systems are not hard to understand, since they are all installed in a logical sequence.

WATER PIPING

The water service pipe is the cold-water line from a water main in the street or a well or other source, such as a spring, a lake, etc., into a building. A valve should be installed on the cold-water pipe at the point where it enters the building. This valve will control all of the water in the building. Learn where this valve is and mark it or tag it so that it can be easily located in an emergency. The cold-water pipe will continue from the point where it enters the building to the fixtures requiring cold-water connections. At the location where the water heater (and softener, if used) is located, a tee will be installed to provide a cold-water supply to the water heater.

The valve on the inlet or cold-water supply to a water heater will control or shut off the cold-water supply to the water heater. Thus, it also *all* of the hot water to the building. In the event of a leak in the hot-water piping or in the hot-water heater, the valve on the cold-water supply to the heater can be shut off, shutting off the hot water but leaving the building supplied with cold water.

DRAINAGE PIPING

Fig. 1-1 shows a typical single story and basement home with the soil and waste piping and vent piping.

The drainage system consists of the building drain and the waste and vent stacks with branches. The building drain is that part of the lowest horizontal piping of the sanitary drainage system inside the walls of the building that receives the discharge from soil or waste stacks or branches and conveys it to a point outside the building, where the building drain connects to the building sewer. This point may be three, five, or ten feet outside the building, depending on local codes.

In Fig. 1-1(A) that part of the drainage system inside the heavy wavy line is the building drain. In this case,

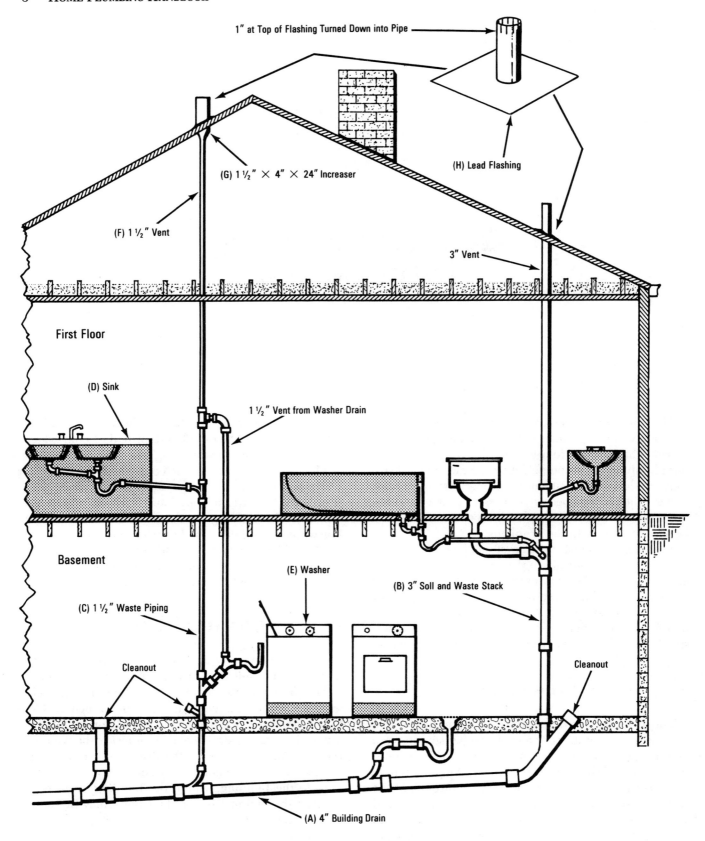

1" at Top of Flashing Turned Down into Pipe

(H) Lead Flashing

(G) 1 ½" × 4" × 24" Increaser

(F) 1 ½" Vent

3" Vent

First Floor

(D) Sink

1 ½" Vent from Washer Drain

Basement

(E) Washer

(B) 3" Soil and Waste Stack

(C) 1 ½" Waste Piping

Cleanout

Cleanout

(A) 4" Building Drain

Fig. 1-1. Plumbing system in a single-story and basement home.

it is under the basement floor. As shown at Fig. 1-1(B), the soil and waste stack receives the discharge from a water closet, bathtub, and lavatory and conveys it to the building drain. A sanitary tee with 1½" or 2" tappings is used to connect the bath drain to the waste stack, ensuring proper venting of the bath drain.

The branch waste line shown at Fig. 1-1(C) receives the discharge from the automatic washer in the basement and the kitchen sink on the first floor. The piping extending above the drain connections at the sink (D) and at the washer (E) is the vent piping. The water closet and the bathtub are vented by a *wet vent*. This is a vent pipe from a water closet or bathtub, or both, which also receives the discharge of a lavatory.

VENT PIPING

Proper venting of soil and waste piping is extremely important. A vent pipe is a continuation of a soil or waste pipe and prevents trap siphonage and back pressure in the soil and waste piping. The main soil and waste stack becomes a vent stack above the waste connection to the highest fixture on the stack. The main vent stack must be continued full size through the roof. The branch vent shown at Fig. 1-1(F) may be extended through the roof, or may be connected to the main vent stack. If the branch vent is extended through the roof in areas where there is danger of frost building up on the inside of the pipe and closing it, an increaser should be used to extend the vent pipe through the roof. The increaser should be 24 inches long so that the change in pipe size is made 12 inches below the roof line. The increaser may be 1½" × 3", 1½" × 4", 2" × 3", or 2" × 4", depending on local codes. A typical increaser is shown at (G) in Fig. 1-1.

At (H) the method of installing a lead roof flashing is shown. The base of the flashing is under the roof shingles; the top of the flashing is turned down into the vent stack. The piping sizes shown on the drawings are intended to serve as a guide, and in general represent minimum acceptable sizes. In many areas, home owners are permitted to do repairs or new work on their own homes, but they are subject to the same requirements of permits, inspections, and tests as the licensed plumber. Local ordinances and codes must be followed.

TOOLS

When you are repairing or installing plumbing, half the battle is won if you have the proper tools to work with. If you plan to do the work yourself, you may at one time or another use most or all of the common plumbing tools shown in Figs. 1-2 to 1-19. You probably already have blade and Phillips-type screwdrivers, pliers, and a hammer or two.

The following list of tools are those which will be needed at some time when doing repair work:

Adjustable wrenches
A 6 ft. rule
Pipe wrenches
Closet auger
Propane tank and burner
Hacksaw
Tubing cutters
Flaring tool
Offset hex wrench
Basin wrench
Plungers
Seat wrench

A 10" pipe wrench and a 14" pipe wrench should be included in a basic tool set for use in the home.

Tools are not cheap, but neither is a plumber's labor. If these tools are purchased as you need them and are well cared for, they will pay for themselves in labor costs saved—and you will still have the tools for the next repair job.

For opening clogged or sluggish drains, where the stoppage is not over 25 feet from the nearest cleanout or fixture opening in the drain line, the spinner-type cable is a good investment. The larger tools shown, such as pipe dies, vises, pipe cutters, soil-pipe tools, etc., can be rented at tool rental shops when and if they are needed, for a fraction of their initial cost.

Slip-joint pliers (Fig. 1-2) is almost indispensable for

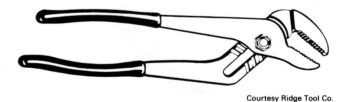

Courtesy Ridge Tool Co.

Fig. 1-2. Slip-joint pliers.

the home-owner-repairman as well as the journeyman plumber. The tool is lightweight and the long handles give extra leverage when gripping pipe, fittings, etc. The teeth are hardened for toughness and long life.

The ball-peen hammer (Fig. 1-3) is the all-purpose hammer for plumbing work. The 12 oz. and 16 oz. sizes in these hammers are the most useful.

The adjustable smooth-jaw wrench (Fig. 1-4) is used to grip and turn squared surfaces of valves, nuts, and fittings.

Courtesy Ridge Tool Co.

Fig. 1-3. Ball peen hammer.

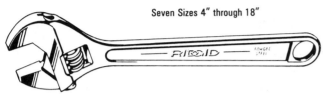

Seven Sizes 4″ through 18″

Courtesy Ridge Tool Co.

Fig. 1-4. Adjustable wrench.

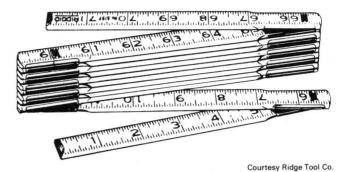

Courtesy Ridge Tool Co.

Fig. 1-5. A 6′ folding rule.

A 6-ft. inside reading rule (Fig. 1-5) is very useful when measuring small repair jobs.

Pipe wrenches (Fig. 1-6) are used to grip pipe and fittings when installing and removing piping. Pipe wrenches have two sets of jaws; each jaw has teeth designed to grip the pipe as pressure is applied to the wrench handle. The upper or hook jaw is adjustable; the lower or heel jaw is fixed. When installing or removing piping, two wrenches are often needed, one wrench to tighten or loosen a pipe or fitting, the other

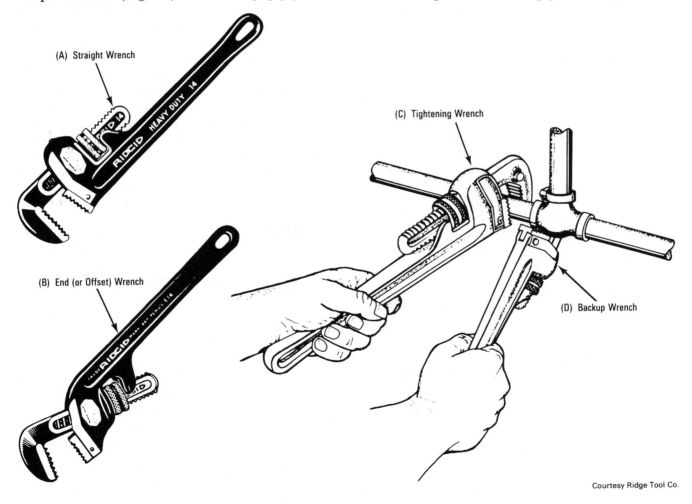

(A) Straight Wrench

(B) End (or Offset) Wrench

(C) Tightening Wrench

(D) Backup Wrench

Courtesy Ridge Tool Co.

Fig. 1-6. Pipe wrenches.

wrench as a backup. When used this way, the wrenches will face in opposite directions (Fig. 1-6 C-D). A backup wrench prevents the piping already installed from being turned when other piping or fittings are added or removed. When using two wrenches, a backup wrench can usually be one size smaller because the piping being held with the backup wrench is already tightened.

End type (or offset) pipe wrenches (Fig. 1-6B) are made for use in tight places: pipes against a wall, or pipes that cannot be turned with the straight pipe wrench. The closet auger (Fig. 1-7) is used to remove or force out a foreign object from a water closet bowl.

Propane torches (Fig. 1-8) using replaceable fuel cylinders are useful for many purposes. A pencil-type burner is used to solder copper tubing and fittings.

Courtesy Ridge Tool Co.

Fig. 1-7. A closet auger.

Fig. 1-8. Propane torch.

A faucet reseating tool (Fig. 1-9) is available with a large range of cutters. This is a special tool designed to do one job, and do it well. See chapter on **Repairing and Replacing Faucets.**

Fig. 1-9. Faucet reseating tool.

Pipe taps (Fig. 1-10) are used to cut female threads, for cleaning rust or corrosion from female threads, and for straightening out damaged threads.

The hacksaw (Fig. 1-11) is a metal cutting saw. Coarse-toothed blades (18 teeth per in.) are used for cutting steel bars, pipe, or heavy metals. Medium blades (24 teeth per in.) are for pipe or metal up to $\frac{1}{8}$ in. thickness. Fine blades (32 teeth per in.) are used for thin metals, thin tubing, etc.

A tubing cutter (Fig. 1-12) is used to cut copper, brass, or aluminum tubing and thin wall conduit. Two sizes cover normal household repair requirements. No. 10, $\frac{1}{4}$ in. through 1 in.; No. 20, $\frac{5}{8}$ in. through $2\frac{1}{8}$ in. All sizes are O.D. (outside diameter of tubing).

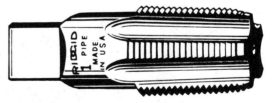

Courtesy Ridge Tool Co.

Fig. 1-10. Pipe tap.

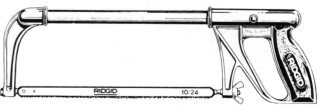

Courtesy Ridge Tool Co.

Fig. 1-11. Hacksaw.

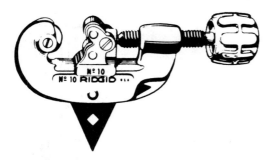

Courtesy Ridge Tool Co.

Fig. 1-12. A tubing cutter.

It is often necessary to thread piping in place. Ratchet-type dies (Fig. 1-13) can be used for this purpose. When needed, they can be rented from tool rental companies.

The flaring tool is used to flare copper, aluminum, and steel tubing for use with compression flare-type fittings. One type of flaring tool is shown in Fig. 1-14.

Courtesy Ridge Tool Co.

Fig. 1-13. Ratchet-type pipe dies.

Courtesy Ridge Tool Co.

Fig. 1-14. Flaring tool.

The offset hex wrench with smooth jaws (Fig. 1-15) will not mar the finish on chrome-plated nuts, fittings, etc.

The basin wrench (Fig. 1-16) is used primarily for removing and installing sink faucets and supplies. The jaws are hinged to turn 180° for either tightening or removing nuts, the hook jaw is spring loaded to close on a nut or fitting. The shank telescopes to desired lengths from 10 in. to 17 in.

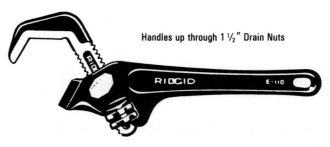

Handles up through 1 ½″ Drain Nuts

Courtesy Ridge Tool Co.

Fig. 1-15. The offset hex wrench.

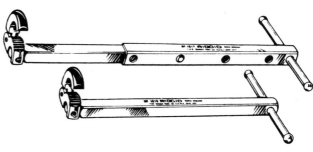

Courtesy Ridge Tool Co.

Fig. 1-16. A basin wrench.

The hand spinner cable (Fig. 1-17) is used to open plugged or sluggish drain piping.

Plungers (or force cups) are used to open clogged drains. Plungers (Fig. 1-18) are made in several sizes; the large bell-shaped plunger is used to open a clogged water closet bowl.

A seat wrench (Fig. 1-19) is used to remove and replace bibb washer seats.

PLUMBING VALVES AND THEIR USES

Each type of valve is designed to be used for a specific purpose (Fig. 1-20). Basically valves serve four purposes:

1. A flow is turned on
2. The flow is throttled or regulated

Fig. 1-17. A hand spinner-type drain cable.

Fig. 1-18. Two types of plungers.

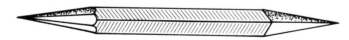

Fig. 1-19. A seat wrench.

3. The flow is turned off
4. Check valves permit flow only in one direction.

Gate valves should be used only as turn-on or stop valves. When the gate is raised and the valve is opened, a straight through full-on flow is permitted—there is virtually no restriction to full flow. Gate valves are not made for throttling or regulating use and are not intended for frequent operation.

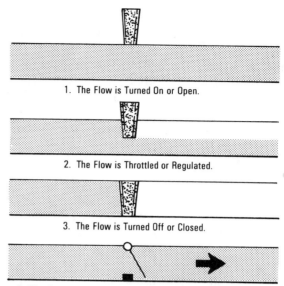

1. The Flow is Turned On or Open.

2. The Flow is Throttled or Regulated.

3. The Flow is Turned Off or Closed.

4. The Flow is Permitted in Only One Direction by Check Valves.

Fig. 1-20. Four functions of specific valves.

Globe valves are very useful as control valves. They have good throttling characteristics. A globe valve operates with fewer turns of the hand wheel and shorter disc travel than a gate valve. Variations of globe valves, called compression stops, are most commonly used around the home. Full flow of water is not usually necessary in the home and the compression stop is much less expensive to manufacture than the gate valve.

The homeowner should be familiar with the locations and functions of the valves in the home. Three or four valves are of special importance. One of them is not a plumbing valve, but deserves mention.

The Main Water Valve

This valve is usually located at the point where the cold-water pipe enters the home. If the water meter is located in the house, the valve should be on the inlet (or street) side of the meter. If the meter is outside, there should be a main shut-off valve at the point where the piping enters the home, or the nearest accessible location. Since this valve will control all of the water in the home, both hot and cold, its location should be known and tagged for use in an emergency.

Hot-Water Valve

This valve controls all of the hot water to the home. In the event of a leak in the water heater, a broken or leaking hot-water pipe, or a hot-water faucet which leaks badly, shutting this valve will shut off the hot water to the home but will still leave the cold water on.

Main Gas and/or Oil Valve

Fire or a leaking pipe could require the immediate shut-off of either or both of these valves. If the gas supply is city gas, the shut-off valve should be at the meter location. If it is bottled gas, the shut-off should be on the piping close to the tank. If heating oil is used, the shut-off valve should be on the oil line at the tank, or at the point where the oil piping enters the home.

Main Electrical Disconnect Switch

This is a plumbing book, true, but the main electrical disconnect switch is a valve and in an emergency, such as trouble with an electric water heater or severe electrical problems, everyone should know how to locate and turn off the main electric disconnect switch at the main electric panel.

Most city and state plumbing codes require the installation of shut-off valves on the piping to every plumbing fixture. These valves should be exposed underneath each lavatory, toilet, and sink. The valves on the hot- and cold-water piping to bathtubs should be behind an access panel built into a wall at the low (or drain) end of the tub. The valves on a built-in shower stall can be under the floor (in a home with a basement) or can be built into the shower valve. The installation of these valves is important; in the event of trouble with any one plumbing fixture, it should be possible to shut off the water to that particular fixture without affecting the rest of the fixtures.

Stop and Waste Valves

Stop and waste valves (Fig. 1-21) are globe valves with a drain feature built into them. Automatic stop and waste valves are made to drain automatically on the downstream (or no pressure side) when the water is turned off. Button-type stop and waste valves will drain when the valve is turned off and the button or cap on the side is opened. Stop and waste valves are normally located inside a heated or protected area and are used to drain piping which could freeze, such as a regular sillcock.

Sillcocks

A common sillcock (Fig. 1-22) is a compression stop valve with a hose thread. Water is normally present at the valve and a common sillcock is subject to freezing unless a stop valve is located in the piping to the sillcock. The stop valve must be shut off and the sillcock opened and drained in cold weather to prevent freezing. Antifreeze-type sillcocks are made with a long body; the shut-off point is inside the home, or in a protected area (Fig. 1-23). Since there is no water in

Courtesy Nibco, Inc.

Fig. 1-21. Stop and waste valves.

Courtesy Nibco, Inc.

Fig. 1-22. Angle sillcock.

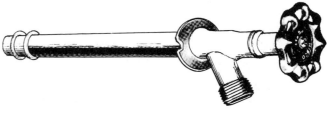

Courtesy Nibco, Inc.

Fig. 1-23. Frost-proof sillcock.

the sillcock when it is shut off, there is no danger of the sillcock freezing and bursting.

There is an important point to remember with either type of sillcock: *Do not leave a hose connected to a sillcock in freezing weather.* A garden hose can hold water up in either type of sillcock and the water in the sillcock or the pipe to the sillcock can freeze and burst the pipe or the sillcock.

CHAPTER 2

Repairing and Replacing Faucets

Many homes have faucets that leak or need only minor repairs to restore them to like-new operating condition. Repair instructions and parts lists are invaluable aids for the do-it-yourself repairman. For this reason, wherever possible I have included photos, model numbers, parts lists, exploded illustrations, and repair and maintenance procedures furnished by manufacturers to help the reader identify a problem.

Repair parts should be readily available from stores selling plumbing supplies for all the brand-name faucets shown in this chapter, as well as for those faucets bearing no brand name.

Miscellaneous repair parts, stems, O-rings, faucet seats, etc., are shown full size in the Appendix.

Basic construction of a faucet is shown in Fig. 2-1 to enable you to disassemble and repair or replace it if needed. Specific instructions are shown in this chapter on sink, lavatory and bathtub faucets. In order to repair almost any faucet, the handle must first be removed from the stem.

The outside, visible parts of most faucets—the handles, escutcheons, flanges, skirts, and stems—are chromium-plated. The makers of good quality faucets use brass for the base metals of these parts, and apply chromium plating to these brass parts for a quality finished product. The makers of cheaper, competitive faucets use "pot" metal as a base metal for handles, escutcheons, flanges, and skirts. Pot metal, which is an alloy of zinc and other metals, is cheaper than brass. Pot metal will corrode when in contact with dissimilar metals. Thus a pot metal handle, mounted on a brass stem, or a pot metal flange screwed on to a faucet body will often be very hard to remove. The corrosion between these parts is occasionally so severe that an old faucet cannot be taken apart by normal methods—it must literally be cut apart using a chisel or a hacksaw. A faucet in this condition is not repairable and must be replaced with a new unit. When a pot metal handle is corroded on a stem, it may be necessary to insert a screwdriver between the bottom of the handle and the flange to pry the handle up in order to get at the stem. This usually causes the bottom of the handle to crumble and break.

The basic steps for repairing any faucet are:

1. Remove the handle. Fig. 2-2 illustrates different methods by which handles are mounted on stems.
2. Remove the stem nut or packing nut—turn the stem nut, or the packing nut counterclockwise to loosen and remove it.
3. Remove the stem—turn the stem in the direction of opening the faucet (turning on the water) to loosen and remove the stem.
4. If the faucet leaks around the stem—replace or add to the graphited packing, or replace the O-rings.
5. If the faucet drips—inspect the bibb washer retainer and replace it if it is damaged.

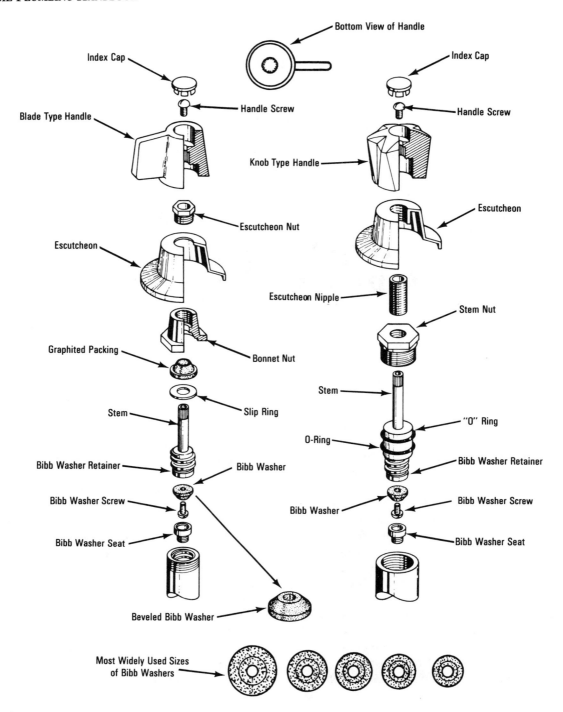

Fig. 2-1. Basic construction of a washer-type faucet.

Turn the bibb washer screw counterclockwise to unscrew and remove it. Install a new bibb washer and replace the bibb screw. Use a soft rubber bibb washer and make certain it is the correct size. The correct size bibb washer will fit easily, without forcing, into the bibb washer retainer.

When the O-ring on a faucet stem becomes worn, water will leak around the stem. To repair this type leak, the O-ring must be replaced. The replacement O-ring must be the same size as the original O-ring. If your faucet has a brand name on it, consult the yellow pages of your phone book under "Plumbing Fixtures and Supplies" for a dealer handling this brand of merchandise, for O-rings or any other repair parts for brand-name products. When reassembling a faucet stem using O-rings, apply a thin coating of waterproof

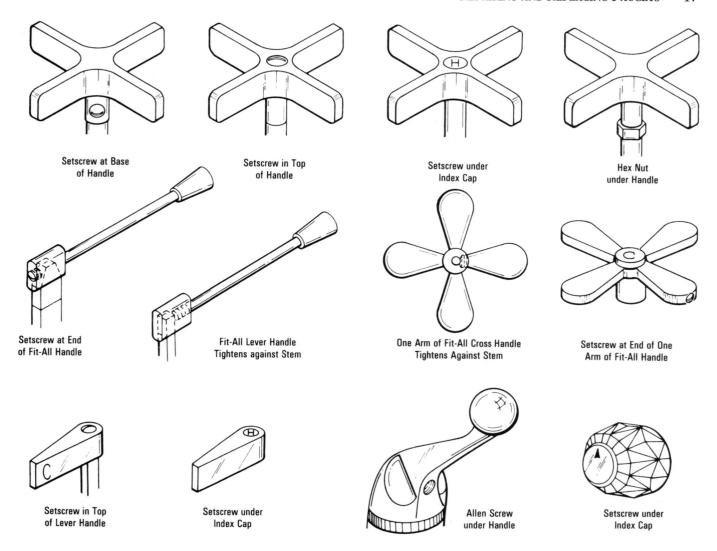

Setscrew at Base
of Handle

Setscrew in Top
of Handle

Setscrew under
Index Cap

Hex Nut
under Handle

Setscrew at End
of Fit-All Handle

Fit-All Lever Handle
Tightens against Stem

One Arm of Fit-All Cross Handle
Tightens Against Stem

Setscrew at End of One
Arm of Fit-All Handle

Setscrew in Top
of Lever Handle

Setscrew under
Index Cap

Allen Screw
under Handle

Setscrew under
Index Cap

Fig. 2-2. Various methods of attaching faucet handles.

grease to the O-rings. To aid in selecting the correct O-ring, full-size illustrations are shown in the Appendix at the end of this book.

Index caps are so called because they indicate the temperature of the water, either hot or cold. On older types of china faucet handles they were screwed into the top of the handle, and tightened with a right-hand (clockwise) thread. Chrome-plated index caps now in use are pressed into the handle and can be removed by inserting a small, thin screwdriver blade under the edge of the cap and prying the cap out. After removing the screw that holds the handle, the handle should lift off; if it does not, tap the underside of the handle until it loosens.

If a faucet handle becomes worn and turns on the stem, a fit-all handle can be used as a replacement. Fit-all handles can be bought in the plumbing department of most hardware stores.

WALL-MOUNTED SINK FAUCETS: FAUCETS BEARING NO BRAND NAME—FIG. 2-3

Turn off the hot- and cold-water supply to the faucet, either under the sink or at the main shut-off valve, before starting repair work. A steady dripping from the spout indicates a replacement of the hot or cold (or both) bibb washers is needed. With the hot and cold stems removed, the bibb washer seats can be inspected. Remove the screws securing the handles to the stems. Fig. 2-2 shows different methods of attaching the handles. It may be necessary to tap lightly on the underside of the handles to remove them. Use an adjustable, smooth-jaw wrench (Fig. 1-4) to turn the bonnet nut counterclockwise.

When the bonnet nuts are free from the faucet body, set the handles back on the stems and turn the stems in

the direction of opening the faucet. When the stems are unscrewed from the faucet body, they can be lifted out. Slide the bonnet nuts off the stem. The stems can then be compared with the full-size stem illustrations to aid in determining which size and type of bibb washers are needed. In a retainer or retainer cup, the bibb washers are held in place by a bibb screw. Some stems have the retainer made onto the stem, others have a separate retainer. If the retainer is damaged or corroded or one side is broken or missing, the bibb washer will not be held in place. If the retainer is the loose type, it can be replaced. If the retainer is damaged but is part of the stem, it will be necessary to replace the stem. The bibb seat should be inspected; it may be necessary to shine the beam from a flashlight into the faucet to inspect the seats. If the seats are chipped or pitted, they should either be replaced or reseated—follow instructions at the end of this chapter on **How to Reseat a Faucet.**

When the stem has been correctly matched, using the stem illustrations, the correct size and number of the bibb washer seat can be purchased. If there is a leak around the faucet stems, add a single strand of graphited packing, 2″ or 3″ long, above the slip ring, between the brass slip ring and the bonnet packing. If the cone bonnet packing has completely deteriorated, replace it with a new bonnet packing. The faucet can now be reassembled.

If there is a leak at the swing spout, the packing in the spout should be replaced. Matching the stems with the illustrations (shown in the Appendix) should aid in learning the brand name of the faucet. Knowing the faucet name should be helpful in choosing the right spout packing part number.

If there is a soap dish mounted on the top of the spout connection (Fig. 2-3), lift the soap dish off. Use a smooth-jaw adjustable wrench to turn the packing nut counterclockwise. When the packing nut is unscrewed and removed from the faucet, the spout can be pulled up and off of the faucet body. Install new packing and reassemble the spout.

FAUCETS BEARING NO BRAND NAME— FIGS. 2-4 AND 2-5

Turn off the hot- and cold-water supply to the faucet, either under the sink or at the main shut-off valve, before starting repair work.

Figs. 2-4 and 2-5 show the two kinds of deck-type sink faucets. Fig. 2-4 shows a top-mounted type. This type has the body of the faucet concealed under a chrome-plated cover. Fig. 2-5 shows a bottom-mounted faucet. The body of a bottom-mount faucet is mounted below the sink or cabinet top. Many of these types of faucets bear no brand name or identification. The stem illustrations should be of help in securing the proper repair parts.

Turn the screws securing the handles to the stems counterclockwise to loosen and remove the handles. Since the handles are often die-cast metal, corrosion between the die-cast metal and the brass stem may make the handles difficult to remove. Tap the handles lightly on the bottom side or pry up on the skirt at the bottom of the handles to loosen. If prying is necessary, exert very little pressure at each point and work all around the handle until it is loosened. Too much pressure applied when prying up on the skirt may cause the die-cast metal to crumble.

If the faucet has escutcheons, a packing nut may secure the escutcheons to the faucet body. Turn the packing nut counterclockwise to loosen and remove the escutcheons. If there is no visible nut securing the escutcheon, turning it counterclockwise should loosen it so that it may be lifted off. If the stem has a packing nut and a packing gland, (Fig. 2-4A), graphited packing is used as a water seal. If the stem has only a stem nut (Fig. 2-4B), then O-rings are used as a water seal. Use a smooth-jaw adjustable wrench, turn the packing gland or the stem nut counterclockwise to loosen and remove the gland or the nut. If the stem is similar to Fig. 2-4A and has a packing nut under the escutcheon, it is not necessary to loosen the packing nut. When the packing gland or the stem nut has been loosened, the stem can be unscrewed and lifted out of the faucet body.

When the stems have been removed, they can be compared with the full-size illustrations of stems to aid in selecting the proper repair parts. The correct bibb screw, seat washer, and washer seat is shown for each stem. If a stem turns but will not tighten in the faucet body, the stem thread is probably stripped and a new replacement stem should be installed. Inspect the washer retainer. If the retainer is damaged, a new one is needed. The retainer is the same size as the bibb washer. The bibb screw should be turned counterclockwise to loosen and remove it. If the bibb screw is corroded, it should be replaced. Inspect the bibb washer seats. If the seats are pitted or chipped, they should be replaced or reseated. Instructions for reseating a faucet are found later in this chapter on **How to Reseat a Faucet.** If a new seat is to be installed, use a seat wrench shown in Fig. 1-19. Turn the wrench counterclockwise to loosen and remove the worn seat.

If O-rings are used on the stem, new ones should be installed before replacing the stem. Compare the worn O-rings with those shown in the Appendix to aid in

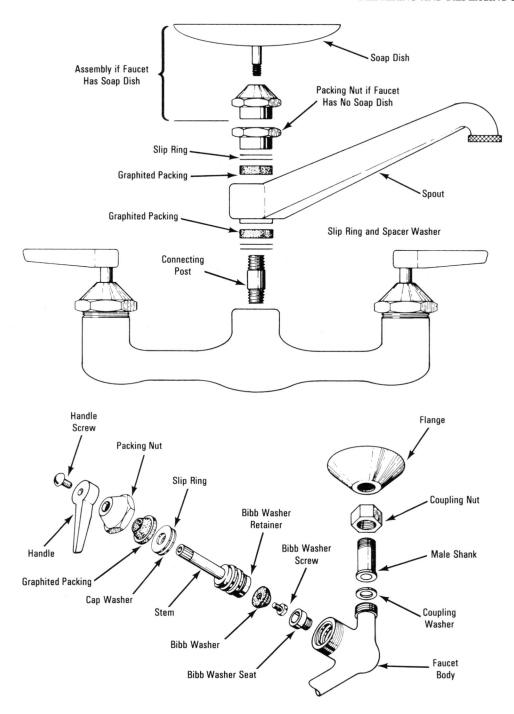

Fig. 2-3. Wall-mount sink faucet.

selecting the correct size. When new O-rings are installed on a stem, they should be coated lightly with a waterproof grease. The grease protects the O-ring when reassembling the faucet and lubricates the O-rings to make the stem turn easily when the faucet is used. If the faucet stems have a packing nut (Fig. 2-4A), the packing nut should be turned clockwise to tighten it slightly when reassembling the faucet. The packing nut should not be too tight; otherwise the stem will be hard to turn. If the faucet leaks around the stem, a new packing should be installed or a single strand (about 2 or 3 in.) of graphited packing can be added to the present packing. Wind the single strand around the stem between the slip ring and the worn packing, and tighten the packing nut. Reassemble the faucet.

If the spout is secured to the faucet by a packing nut

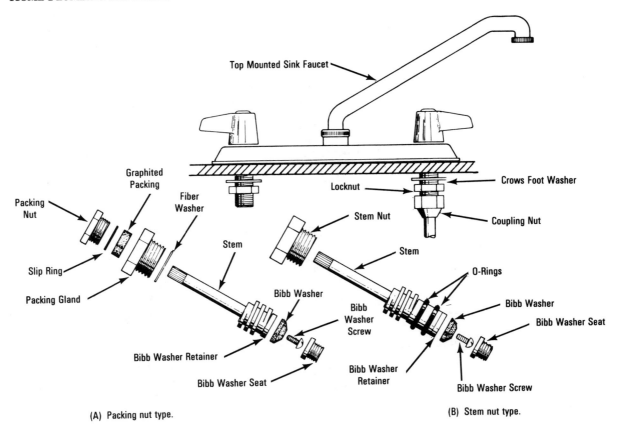

Fig. 2-4. Top-mount deck-type sink faucet.

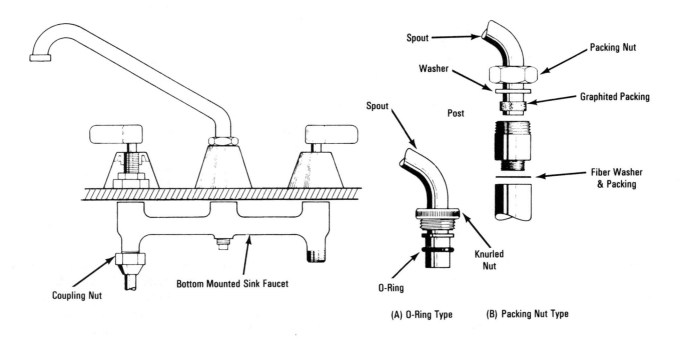

Fig. 2-5. Bottom-mount deck-type sink faucet.

(Fig. 2-5B), turning the packing nut clockwise one quarter turn will often stop the leak. If the packing is badly worn, new packing should be installed.

If the spout uses an O-ring as a water seal (Fig. 2-5A), the worn O-ring can be compared with the illustrations of O-rings shown in the Appendix to aid in selecting the correct replacement. If you were able to identify the brand name of the faucet by the stem illustrations, you may be able to cross check the O-ring by the number against the list of faucets shown in the O-ring illustration. Coat the O-ring lightly with waterproof grease before reassembling the spout.

REPAIRING LAVATORY FAUCETS—FAUCETS BEARING NO BRAND NAME

Turn off the hot- and cold-water supply to the faucet, either under the sink or at the main shut-off valve, before starting repair work.

If a lavatory faucet that leaks and needs repairing has a name on the faucet, the correct repair parts should not be hard to find. In the Appendix that shows actual size stems, you will see that the stem illustration also lists the kind and size of bibb washer to use, and the correct replacement bibb washer seat. The Appendix shows the full-size drawings of stem packings and of the O-rings and the make of faucets which they fit. There are many faucets that have no brand name on them. Here again, the full-size illustrations of stems will aid in finding the correct repair parts.

To repair the faucet, first remove the handle. See Fig. 2-2, which illustrates different types of handles and the way they are secured to the faucet stem. Many handles are made of die-cast metal and corrosion between the die-cast metal and the brass stem may make the handle difficult to remove. When the screw securing the handle to the stem has been removed, if the the handle cannot be lifted off the stem, tap the underside of the handle lightly with a small hammer. Some types of faucet handles have a skirt at the bottom and it may be necessary to pry up on the skirt with a screwdriver blade to loosen the handle. If prying is necessary, exert *very little pressure* and work all around the skirt, prying just a little at each point. Exerting too much pressure will cause the die-cast metal to crumble and break. If the faucet has no escutcheons, use an adjustable smooth-jaw wrench to turn the packing gland or the stem nut counterclockwise to loosen and remove it. Set the faucet handle back on the stem and turn the handle in the direction of opening the faucet to loosen and remove the stem. When the stem is removed, inspect the seat. (It may be necessary to use a flashlight to do this.) If the seat is pitted or chipped, use a seat wrench (Fig. 1-19) to remove the old worn seat. Or if a seat is not readily obtainable, if you have a reseating tool, such as is shown in Fig. 1-9, the old worn seat can be restored to new condition by following the instructions given in this chapter on **How to Reseat a Faucet.**

If water leaks around the stem when the faucet is turned on, install a new bonnet packing, or add a few strands of graphited packing to the old bonnet packing. Often one strand of packing, 2 or 3 in. long, can be added on top of the brass slip ring. When the faucet is reassembled, the bonnet nut will compress the graphited packing into shape. Quite often when the stem has been removed, the bibb screw will be badly corroded. If it can be removed, a new bibb screw should be installed. If the screw is twisted off, your local plumbing shop may be able to drill and tap the stem for a new screw. The bibb screw must be turned counterclockwise to loosen and remove. When replacing the bibb washer, use the correct size and type. These are shown for each stem in the Appendix. With some types of faucets, the stem is threaded into the body of the faucet. With this type faucet, if the stem turns but will not tighten, the thread on the stem is probably stripped. A new stem will be necessary to correct the problem.

Certain types of single-lever faucets use a cartridge element. It will be necessary to install a new cartridge when faucets drip. When new O-rings are installed on a stem, they should be lightly coated with a waterproof grease. The grease protects the O-rings when reassembling the faucet and lubricates them to make the stem turn easily when the faucet is used.

Pop-up drains consist of a lift rod connected to a horizontal rod. The horizontal rod extends into the pop-up assembly. The drain plug is lifted or lowered by the combined action of these rods. The horizontal rod may have a ball seating against a nylon seat, or it may be a cam-action-type rod and depend on a graphited packing for a water seal. If a leak develops at this point, tightening the nut securing the horizontal rod will sometimes stop the leak. Replacement of the pop-up assembly is the recommended cure for this problem (see Fig. 2-6).

The following problem and cure examples are included on some of the most popular model faucets and valves. These examples are shown as a guide only in the disassembly sequence and general overall repair. Although your particular valve or faucet may not be included in these examples, the problem, cure, and disassembly procedures are generally the same.

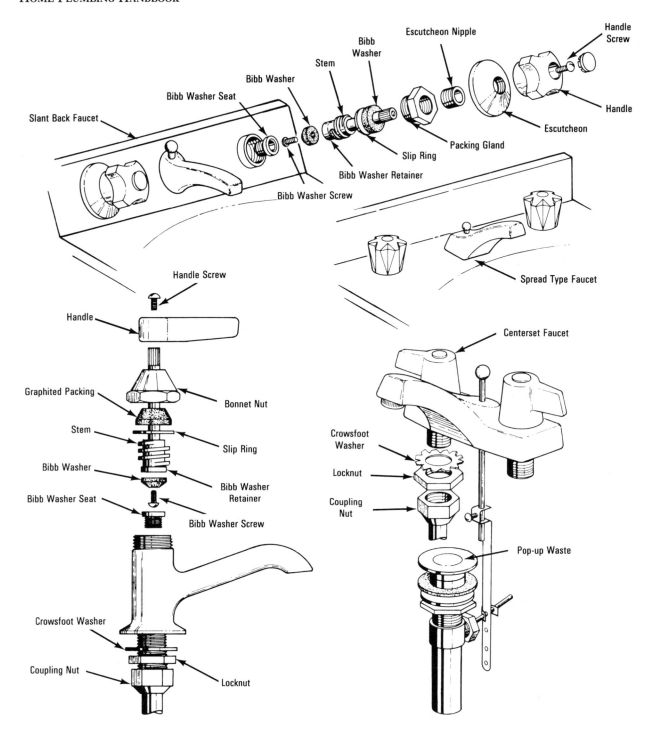

Fig. 2-6. Faucet and pop-up drain assemblies.

American Standard Single-Lever Faucet

R-4150-1 less hose and spray
R-4155-1 with hose and spray

PROBLEM: Faucet drips—will not shut off completely (Fig. 2-7).
CURE: Replace parts 6-7-8-9 (2 each). Use repair kit No. 72495-07. The kit also contains item 3 (O-ring).

The spout (1) must be removed and the escutcheon (4) removed. Turn the spout nut (1a) counterclockwise to loosen. When the nut is loose, lift off the spout. Lift off the escutcheon (4) to expose the faucet body. Turn the two plugs (10) counterclockwise to loosen, then remove. Remove the worn parts 6-7-8-9. Take one each of 6-7-8-9 from the repair kit and install them in

Single Control Sink Fittings

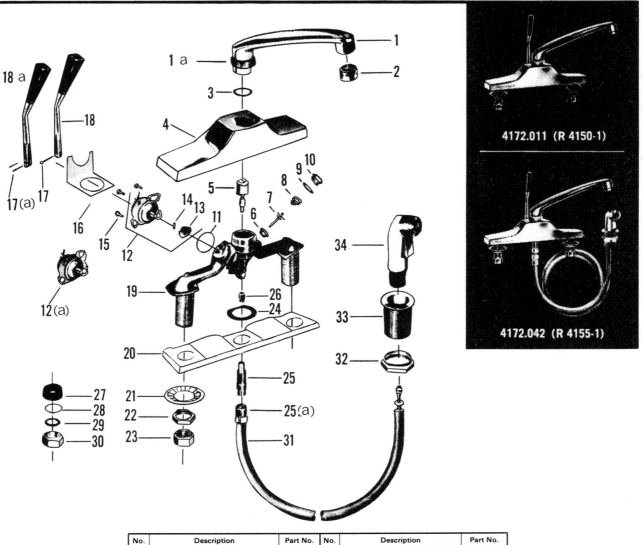

4172.011 (R 4150-1)

4172.042 (R 4155-1)

No.	Description	Part No.	No.	Description	Part No.
1	Spout S/A w/Aerator & "O" Ring	61011-02	19	Body S/A W/Hyseal Valve Seat	61056-07●
2	Aerator	56135-02	20	Faucet Mounting Gasket	61013-07
3	"O" Ring	55234-07	21	Shank Washer	754-17
4	Escutcheon	61006-02	22	Jamb Nut	300-27
5	Diverter Valve	12435-07	23	Swivel Nut	24220-07
6	Hyseal Valve Seat	12002-07	24	Center Washer	61014-07●
7	Hyseal Valve Stem	61428-07	25	Tube Connector	63418-07
8	Conical Spring	72510-07	26	Pipe Plug	56519-09●
9	Strainer Plug Gasket	12007-07	27	Slip Joint Gasket	1010-17
10	Strainer Plug	12268-07	28	Slip Joint Washer	21391-07
11	Control Mounting Gasket	12035-07	29	Slip Joint Washer	21545-07
12	Control S/A w/Parts 13 & 14	12375-04	30	Slip Joint Nut	24906-07●
12A	Control Sub Assembly	61226-04	31	Hose S/A	1178-17
13	Cam	12390-07	32	Lock Nut	61300-06
14	Cam Pin	12034-07	33	Spray Holder	
15	Control Mounting Screws	61020-04	34	Sprayhead S/A	63181-07
16	Rear Closure	12067-07			
17	Handle Screw	12326-07		Sub-assemblies	
17A	Handle Pina (2)	63154-07		Spray & Hose S/A	
18	Control Lever S/A	61051-04		(Parts 31 & 34)	63172-07
18A	Control Lever S/A w/ 2 Pin Openings	61228-04		Control & Lever S/A (Parts 12A-17A-18A)	61232-04

Repair Kit #72495-07 consisting of ((1) Part 3 and (2) each 6, 7, 8, 9).

● NOT AVAILABLE

Courtesy American Standard

Fig. 2-7. American Standard single-lever faucet.

the order shown in the exploded drawing (Fig. 2-7) in the hot (left) side of the faucet body and in the cold (right) side of the faucet body. Replace the two strainer plugs (10). Remove the worn O-ring (3) from the spout and install the new O-ring. Lightly coat the O-ring with waterproof grease to prevent damage when installing the spout. Replace the escutcheon (4) and insert the spout into the faucet body and tighten the spout nut (1a). *Do not overtighten the spout nut.*

PROBLEM: Water leaks around the spout connection to the faucet.

CURE: Replace the O-ring (3), Part No. 55234-07. To replace the O-ring (3) remove spout and replace the worn O-ring, as explained above. Even if the faucet does not drip, it would be wise to get the complete repair kit and replace items 6-7-8-9, since the spout has to be removed just to replace item (3).

PROBLEM: Spray head or spray hose leaks.

CURE: Replace complete hose and spray assembly, Part No. 63172-07. Hose is removed from underneath by turning coupling nut (25a) counterclockwise to loosen and remove hose.

PROBLEM: Spray does not operate properly.

CURE: Remove spout and replace diverter valve, 5, Part No. 12435-07. Grease O-ring as outlined previously.

Repairing American Standard Wall-Mounted Sink Faucet
(R-4213)

PROBLEM: Faucet drips—will not shut off completely (Fig. 2-8).

CURE: Replace seat washer, inspect seat, replace or resurface seat if it is chipped or pitted.

Remove the handle screw and the handle. Use a smooth-jaw wrench and turn the cap (3) counterclockwise to loosen and remove. Set the handle back on the stem, turn the stem in the direction of opening the faucet (counterclockwise on the hot stem, clockwise on the cold stem), and unscrew the stem. Inspect the washer seat (10). If the seat is chipped or pitted, use the seat wrench to turn the seat counterclockwise to loosen and remove. Replace the old seat, if needed, with Part No. 174-14. The worn seat can be resurfaced without removing it, using reseating tool. Install new seat washer (8) and reassemble the faucet.

PROBLEM: Water leaks around the faucet stem.

CURE: Add a strip of graphited packing to present packing (4) or install new stem packing.

Remove the handle and cap as outlined above, and add a strip of graphited packing above present packing or install new packing Part No. 1311-07.

PROBLEM: Water leaks around the spout.

CURE: Install new spout packing.

Use smooth-jaw wrench and turn spout packing nut counterclockwise to loosen. Lift off the soap dish, if used, and lift up and remove the spout. Replace the two worn spout packings, one below the spout and one above, with new packing, Parts No. 473-07. Reassemble the spout, placing the lock washer and space washer above the upper packing as shown.

Kohler Rockford Ledge (Top-Mount) Sink Faucet
(K-7827T with hose and spray)
(K-7825T less hose and spray)

PROBLEM: Faucet leaks—drips (Fig. 2-9).

CURE: Replace the seat washer, inspect the renewable seat (12). If the seat is chipped or pitted, replace the seat.

Remove the screw (1) and the handle (2). Use an adjustable wrench to turn the bonnet (3) counterclockwise to loosen and remove. When the stem (7) is screwed into the plunger (8), the stem and the plunger can be lifted out of the body of the faucet. Lift out the sleeve (11) and inspect the seat. If it is chipped or pitted, replace it with new part (12), Part No. 23004. The seat is pressed into the bottom of the sleeve. To remove the seat, insert a nail set through two opposite holes in the sleeve and tap the square head of the nail set lightly with a small hammer. The worn seat will be forced out of the sleeve. Insert the new seat into the sleeve and tap it lightly with a small hammer to "set" it. Replace the O-rings (4 and 5) on the stem with new O-rings, Part Nos. 34264 and 34263. Lightly coat the O-rings with waterproof grease to prevent damage when reassembling the faucet. Note that the sleeve is grooved to receive the plunger.

PROBLEM: Faucet leaks around the spout.

CURE: Replace the worn O-rings on the spout.

Use an adjustable wrench to turn the coupling nut (16) counterclockwise to loosen and remove. Remove the spout. Replace the two worn O-rings (18), Part No. 40933. Lightly coat the O-rings with waterproof grease to prevent damage when reassembling the spout.

PROBLEM: Spray hose does not work properly.

CURE: Replace the auto spray unit (21), Part No. 39931.

Remove the spout as outlined above. Turn the spout post counterclockwise to loosen and remove the post. Lift out the worn spray unit and install the new unit. Install new O-rings (18) as outlined above, before reassembly of the spout.

Repairing Kohler Exilla Single-Lever Faucet
(K-7840, K-7842 Sink Fitting)

PROBLEM: Faucet drips—will not shut off completely (Fig. 2-10).

Double Faucets
Renewable Seats

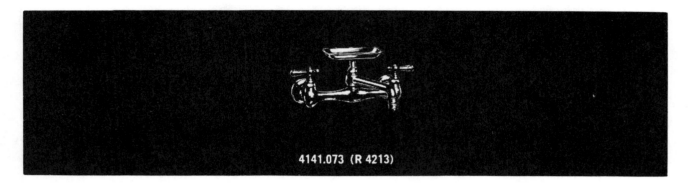

4141.073 (R 4213)

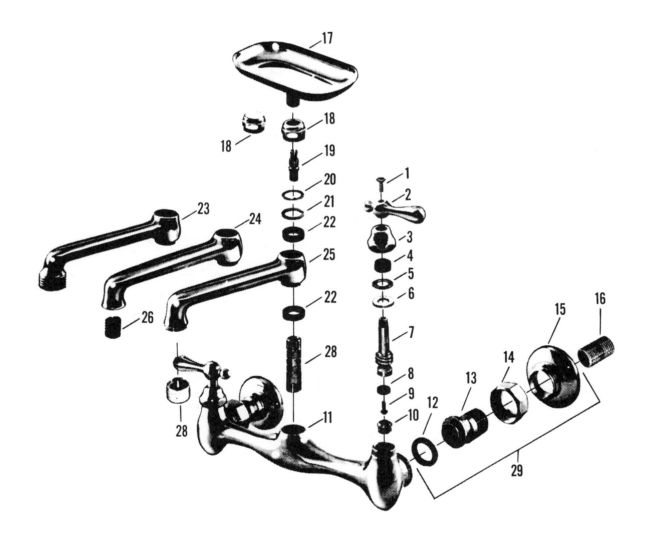

Fig. 2-8. American Standard wall-mount faucet.

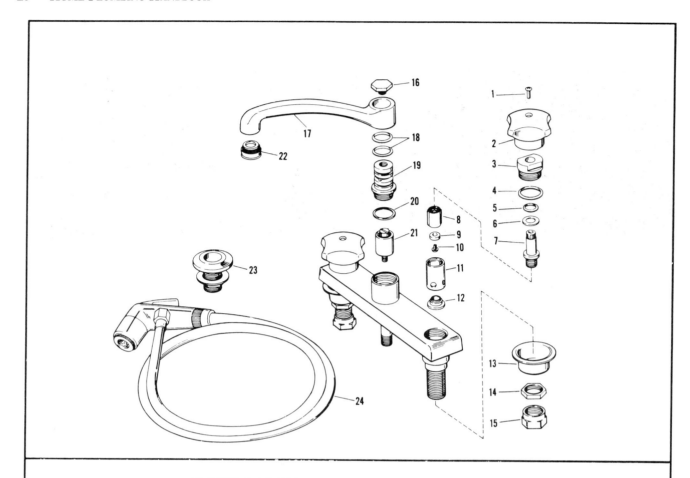

K-7827-T ROCKFORD LEDGE SINK FITTING PARTS
TRITON II SERIES
(K-7825-T Same as K-7827-T Except Loess Hose and Spray)

ITEM NO.	PART NO.	DESCRIPTION	ITEM NO.	PART NO.	DESCRIPTION
1	53448	Screw	14	32010	Lock Nut
2	41980	Handle (Specify Index)	15	32751	Coupling Nut
3	34260	Bonnet	16	34312	Spout Cap Nut
4	34264	"O" Ring	17	38687	Spout w/Aerator
5	34263	"O"Ring	18	40933	"O" Ring
6	34265	Washer	19	38686	Spout Post
7	34320	Stem	20	40950	Gasket
8	22947	Plunger w/Seat Washer & Screw	21	39931	Auto Spray Unit
9	39541	Seat Washer	22	41056	Aerator
10	34848	Screw	23	34329	Hose Guide S.A.
11	34842	Sleeve	24	38613	Hose & Spray
12	23004	Renewable Seat		34570	Spray Only
13	34274	Spacer		38614	Hose Only

Items 5-12 Inclusive, Specify 22932 Valvet for Hot and Cold Valve.

Note: For fitting with soap dish, specify No. 37766 Post and Nut Assembly in place of Item 16. Soap dish for above, specify 37668.

Fig. 2-9. Kohler top-mount sink faucet.

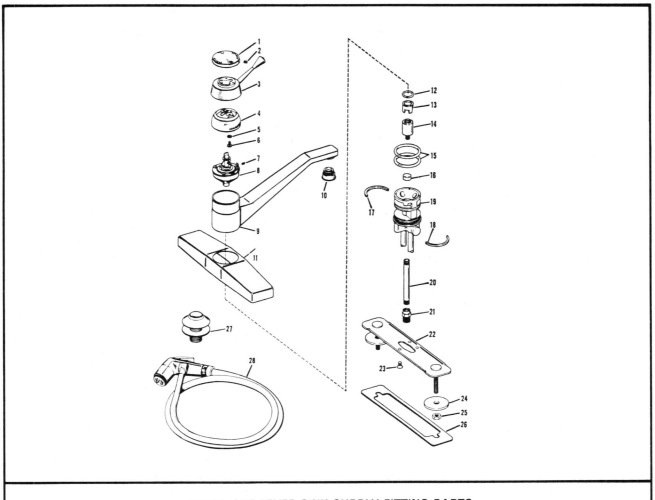

EXILLA ONE LEVER SINK SUPPLY FITTING PARTS
K-7840 and K-7842 Exilla Sink Fitting

Item No.	Part No.	Description
1	38638	Cap
2	35739	Screw
3	38622	Handle & Cover Assembly } #38623 handle, cover & skirt assembly
4	35762	Skirt
5	35789	Washer
6	35744	Screw
7	35804	Set Screw
8	35774	Valve Unit (RC-2)
9	38750	Spout
10	41056	Aerator
11	38616	Shroud
12	32657	'O" Ring
13	38633	Spacer
14	38636	Diverter Unit
15	38618	"O" Ring (2 required)

Item No.	Part No.	Description
16	38701	Screen (2 required)
17	38620	Clamp
18		
19		Valve Body (not available)
20	20667	Connection for Hose & Spray
21	20666	Adaptor
22	38628	Base Plate
23	38635	Screw (3 required)
24	38619	Washer
25	51533	Nut
26	38630	Gasket
27	34329	Hose Guide Assembly
28	38613	Hose & Spray
	38614	Hose only
	34570	Spray only

Fig. 2-10. Kohler single-lever sink faucet.

CURE: Replace the valve unit (item 8, Part No. 35774).

Since it is necessary partially to disassemble the faucet to replace the valve unit, it would also be wise to install new O-rings on the valve body (item 15, Part No. 38618—2 required) and also replace item (12), O-ring (Part No. 32657).

The numbered exploded drawing shows the steps for disassembly of this faucet. Remove the cap, handle, cover, and skirt assembly. Remove the old valve assembly. Lift the spout off and remove the two O-rings (item 15) from the valve body. Remove the old O-ring (item 12), install new O-ring in its place, and install two new O-rings (item 15) on valve body. Lightly coat the outsides of the new O-rings with a waterproof grease. This will prevent damage when reassembling the faucet. Reassemble the faucet. The hose and spray are removed for replacement, if necessary, from underneath the sink. The hose and spray unit are coupled to item (21) by a nut on the hose. Turn this nut counterclockwise to loosen and remove the old hose and spray unit. Replace the old unit with Part No. 38613, hose and spray.

PROBLEM: Hose and spray do not operate properly.
CURE: Disassemble faucet as outlined above. Install new divertor unit, item (14), Part No. 38636.

Repairing Kohler HYKA
Single-Lever Faucet
(K-7830, K-7832 Sink Fitting)

PROBLEM: Faucet drips—will not shut off completely (Fig. 2-11).
CURE: Replace cartridge (item 13, Part No. 39925).
PROBLEM: Water leaks around spout connection to faucet body.
CURE: Replace O-ring (item 3, Part No. 29464).

Since it is necessary to remove the spout and cover in order to replace the cartridge, the old O-ring (item 3) should be replaced.

Use a smooth-jaw wrench and turn nut (2a) counterclockwise to loosen. Lift off the spout and replace the old O-ring (item 3) with a new O-ring, Part No. 29464. Turn the lock ring (item 6) counterclockwise to loosen and remove. Lift off the cover (item 7), loosen and remove the four screws (item 12); the stem and cap assembly can be set aside for reassembly later. Remove the two screws (item 14) and remove the old cartridge (item 13). Install the new cartridge, RC-1, (item 13, Part No. 39925), replace the screws (item 14), reassemble the stem and cap, assemble and replace the cover. Coat the new O-ring, (item 3) lightly with waterproof grease, and reassemble the spout. Do not overtighten the nut on the spout (2a).

The hose and spray assembly (item 25, Part No. 38613) are removed for replacement if necessary from underneath the sink. The hose is connected to the valve body by a nut that should be turned counterclockwise to loosen and remove.

Kohler Clearwater Wall-Mounted
Sink Faucet
(K-7856)

PROBLEM: Faucet leaks—drips (Fig. 2-12).
CURE: Replace the worn seat washers (item 9). Inspect the renewable seats (item 12) and replace them if they are chipped or pitted.

Remove the handle screw (item 1) and the handle. Use an adjustable wrench to turn the bonnet (item 3) counterclockwise to loosen and remove. When the stem (item 7) is screwed into the plunger (item 8), the plunger can be lifted out of the faucet body. Lift the sleeve (item 11) out and inspect the washer seat. The washer seat should be replaced with a new seat, Part No. 23004. The washer seat should be replaced if it is chipped or pitted. The seat is pressed into the bottom of the sleeve. To remove the worn seat, insert nail set through two opposite holes in the bottom of the sleeve and tap the square head of the nail set lightly with a small hammer. The worn seat will be forced out of the sleeve. Insert the new seat into the sleeve and tap it lightly to "set" it.

Replace the worn seat washer (item 9); the two O-rings (items 4 and 5) should also be replaced when installing new seat washers. When reassembling the faucet, note that the sleeve is grooved to receive the plunger. The stems and plungers on the hot-water side of the faucet are different from those on the cold-water side; therefore, it is best to repair and reassemble one side of the faucet before repairing the other side, to avoid mixing the parts. Lightly coat the new O-rings with waterproof grease before reassembly.

PROBLEM: Water leaks around the spout.
CURE: Replace the worn O-rings (items 17 and 18).

Remove the soap dish and turn nut (item 15) counterclockwise to loosen and remove the nut and the post. Remove spout. Remove the two worn O-rings (items 17 and 18) and install two new O-rings, Part No. 40933. Lightly coat the new O-rings with waterproof grease before reassembly.

Repairing Kohler Clearwater
Wall-Mount Sink Faucet
(K-8655-A, K-8656-A, K-8657, K-8659-A)

PROBLEM: Faucet drips—will not shut off completely (Fig. 2-13).
CURE: Replace seat washers (item 8) and install new renewable seats (item 10) or resurface present seats if they are chipped or pitted.

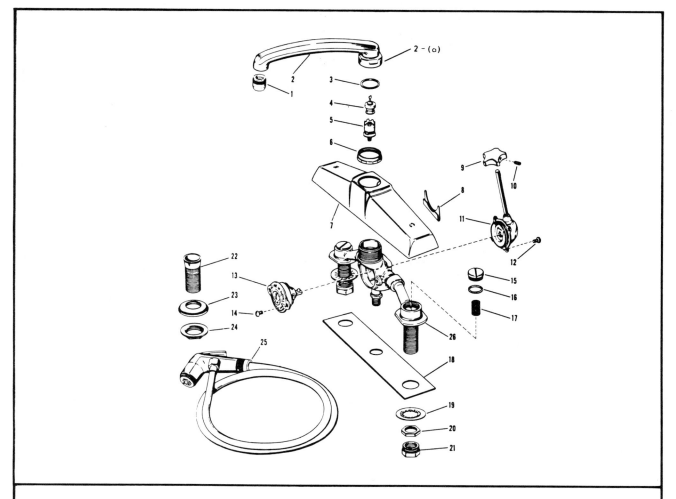

HYKA ONE LEVER SINK SUPPLY FITTING PARTS
K-7830 and K-7832 Hyka Sink Fitting

Item No.	Part No.	Description
1	41056	Aerator
2	41431	Spout w/41434 Coupling Nut and 41433 Spout Post
3	29464	"O" Ring
4		
5	39867	Auto Spray Unit
6	41435	Lock Ring
7	41432	Cover
8	41443	Clip
9	35118	Handle
10	35176	Screw
11	41447	Stem & Cap Assembly
12	35744	Screw (4 required)
13	39925	Cartridge (RC-1)
14	49931	Screw (2 required)
15	41437	Cap

Item No.	Part No.	Description
16	34300	"O" Ring
17		Screen (not available)
18		Gasket (not available)
19	33420	Washer
20	32010	Lock Nut
21	32751	Coupling Nut
22		
23	34329	Hose Guide Assembly
24		
25	38613	Hose and Spray
	38614	Hose only
	34570	Spray only
26	41449	Valve Body w/Shanks for K-7832
	41441	Valve Body w/Shanks & 40630 Plug for K-7830

Fig. 2-11. Kohler single-lever sink faucet.

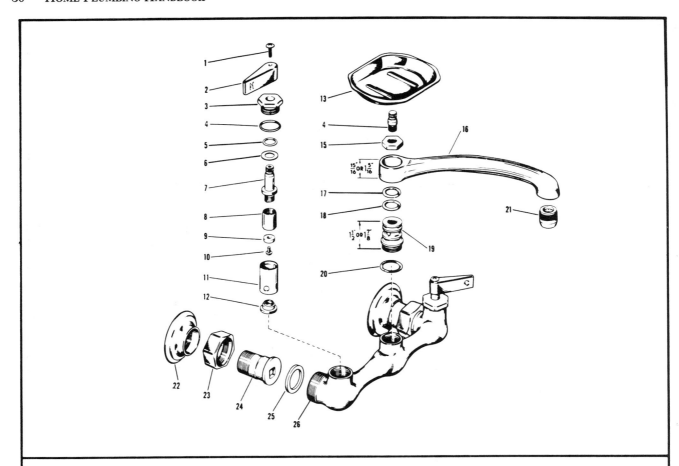

K-7856 CLEARWATER SINK FITTING PARTS

Item No.	Part No.	Description
1	33357	Screw
2	34064	Handle (specify index)
3	34365	Bonnet
4	34264	"O" Ring
5	34263	"O" Ring
6	34265	Washer
7	34319	Cold Stem
	34320	Hot Stem
8	22947	Cold Plunger w/seat washer & screw
	22948	Hot Plunger w/seat washer & screw
9	39541	Seat Washer
10	34848	Screw
11	34842	Sleeve
12	23004	Renewable Seat
13	37668	Soap Dish

Item No.	Part No.	Description
14	34532	Post 37766 Post & Nut
15	34533	Nut Sub-Assembly
16	37750	Spout with Aerator—$^{15}/_{16}$"
	38687	Spout with Aerator—1 $^5/_{16}$"
17	40933	"O" Ring
18	40933	"O" Ring
19	34430	Spout Post—1 $^1/_2$"
	38685	Spout Post—1 $^7/_8$"
20	32657	"O" Ring
21	41056	Aerator
22	39708	Flange
23	40748	Nut
24	40599	Shank inside Thread
	40598	Shank outside Thread
25	40718	Washer
26	34512	Body

Note: Valve parts for K-7857 and K-7856 same as above.
Items 5 through 12, specify 22917 Valvet for cold valve,
22932 Valvet for hot valve.

Fig. 2-12. Kohler wall-mount sink faucet.

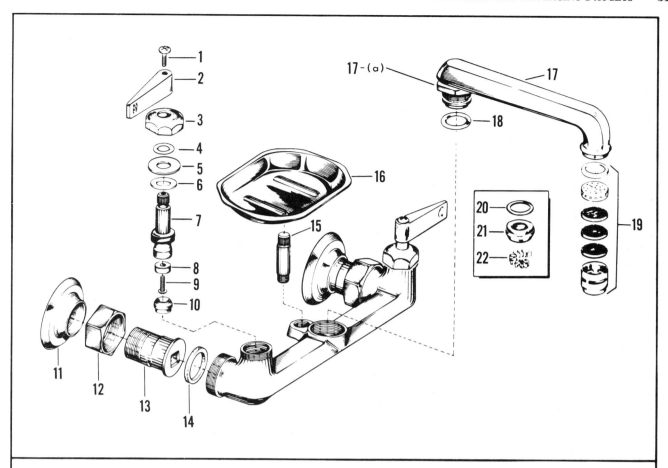

CLEARWATER SINK FITTING PARTS
For K-8655-A, K-8656-A, K-8657, K-8659-A

Item No.	Part No.	Description
1	33357	Screw
2	34064	Handle (specify index)
3	34074	Bonnet
4	34117	"O" Ring
5	39881	Gasket
6	39540	Washer
7	31871	Hot Stem with Washer and Screw
	31872	Cold Stem with Washer and Screw
8	39541	Seat Washer
9	31490	Screw
10	40602	Renewable Seat
11	39708	Flange
12	40748	Coupling Nut
13	40599	Shank—Female Connection
	40598	Shank—Male Connection
14	40718	Gasket

Item No.	Part No.	Description
15	39888	Post
16	37668	Soap Dish
17	37700	5" Spout with Streambreaker for K-8655 or K-8656
	37702	9" Spout with Streambreaker for K-8659
	37701	5" Spout with Hose Connection for K8657
	37692	5" Spout with Aerator for K-8659-A
18	32725	"O" Ring
19	37725	Aerator

Fig. 2-13. Kohler wall-mount sink faucet.

Remove screw (item 1) and handle (item 2). Use a smooth-jaw wrench to turn the bonnet nut (item 3) counterclockwise to loosen and remove. Set the handle back on the stem and turn the stem in the direction of opening the faucet until the stem is loose; lift the stem out. Inspect the seat; if it is pitted or chipped, use a seat wrench to turn it counterclockwise to loosen and remove. The seat may be resurfaced without removing if a new seat is not available.

Remove the old seat washer (item 8) and install a new seat washer, Part No. 39541. Remove the old O-ring (item 4) from the stem and replace with a new ring, Part No. 34117. Lightly coat the new O-ring with waterproof grease before reassembly. Insert the stem into the faucet body and reassemble bonnet and handle. The above directions are for repair of the hot-water side of the faucet; the cold-water side is repaired in the same manner.

PROBLEM: Water leaks around spout connection to faucet body.
CURE: Replace the O-ring (item 18).

If the spout leaks at the connection to the faucet body, use a smooth-jaw wrench to turn the nut (17a) counterclockwise to loosen the nut. When the nut is free, the spout can be lifted out. Remove the old O-ring (item 18) and replace it with a new ring, Part No. 32725. Lightly coat the new O-ring with waterproof grease before reassembling the spout. Do not overtighten the spout nut (item 17a).

American Standard Centerset Lavatory Faucet

N-2001, with pop-up waste
N-2041, with chain and plug waste
PROBLEM: Faucet leaks—drips (Fig. 2-14).
CURE: Replace aquaseal diaphragms. If the washer seats are chipped or pitted, replace or resurface the washer seats.

Insert a small screwdriver blade under the edge of the index cap (1) and lift. Remove the handle screw (2) and the handle. Use an adjustable wrench to turn the lock nut counterclockwise to loosen and remove the lock nut. Lift the stem (6) and the worn aquaseal diaphragm out. Inspect the seat (10). If the seat is chipped or pitted, replace it with a new seat, Part No. 862-14. Remove the damaged seat with the seat wrench.

The damaged seat can be resurfaced in place, using the reseating tool.

Install new aquaseal diaphragms (Part No. 72960-07) on the ends of the stems and reassemble the faucet.

American Standard Single-Lever Lavatory Faucet

N-2055, with pop-up waste.
PROBLEM: Faucet leaks—drips (Fig. 2-15).
CURE: Replace parts 7, 8, 9, 10, using repair kit No. 12417-07.

No. 1, bolt, is not used on this faucet. Turn the control knob (2) counterclockwise to unscrew it from the waste lever. Turn the waste control bushing (3) counterclockwise to loosen and remove it. Remove the escutcheon (4). Use a screwdriver to turn the valve plug (6) counterclockwise to remove the plug. Remove worn parts 7, 8, 9, 10. The repair kit has two each of parts 7, 8, 9, 10. Install one new set of these parts in each side of the faucet and reassemble the faucet.

American Standard (8") Spread Lavatory Faucet

PROBLEM: Faucet leaks—drips.
CURE: Replace the aquaseal diaphragms; if the washer seats are chipped or pitted, replace or resurface the washer seats.

The repair procedure is the same for this faucet as for the N-2001 or N-2041 faucets. See instructions for these faucets. The part numbers for the aquaseal diaphragms and the seats are also the same as for the N-2001 and N-2041 faucets.

Kohler Constellation and Galaxy Centerset Lavatory Faucets

(K-7400, K-6960)
PROBLEM: Faucet drips (Fig. 2-16).
CURE: Replace the worn seat washers, item 9; inspect and replace the renewable seats if they are chipped or pitted.

Remove the handle screw (item 1) and the handle. Use an adjustable wrench to turn the bonnet (item 3) counterclockwise to loosen and remove. When the stem (item 7) is screwed into the plunger (item 8), the plunger can be lifted out of the faucet body. Lift the sleeve (item 11) out and inspect the washer seat. The washer seat should be replaced if it is chipped or pitted. The seat is pressed into the bottom of the sleeve. To remove the worn seat, insert a nail set through two opposite holes in the bottom of the sleeve and tap the square head of the nail set lightly with a small hammer. The worn seat will be forced out of the sleeve. Insert the new seat (item 12), Part No. 23004, into the sleeve and tap it lightly to "set" it. Replace the worn seat washers (item 9). The two O-rings (items 4 and 5),

Centerset Lavatory Fittings
Heritage Trim—Aquaseal

2101.012 (N 2001) | 2104.016 (N 2041)

no.	description	part no.
1	Button - Specify Index	1234-22
2	Handle Screw	915-17
3	Handle	5358-02●
4	Lock Nut	855-17
5	Stem Nut	72956-07
6	Stem w/swivel & Friction Ring	72952-07●
7	Friction Ring	246-37
8	Stop Ring	72959-07
9	Aquaseal Diaphragm	72960-07
10	Seat	862-14
11	Lift Rod w/knob	4358-02
12	Lift Rod Guide	25845-02●
13	Body w/seats, Shanks & Rod Guide	8176-02●
	Body w/seats & Shanks	8177-02●
	Body w/seats & Shanks	8078-02●
14	End Trim	279-12
15	Aerator	56135-02
16	Spray Head S/A	5450-02●
17	Sleeve for Chain	275-04
18	Chainstay Nut	28182-04
19	Chain	1112-02
20	Link	1128-22
21	Rubber Stopper	1012-12
22	Friction Washer	54538-09
23	Lock Nut	300-27
24	Coupling Nut	24220-07
25	Assembled Handle Specify Index	6458-02
26	Aquaseal Trim Less Lock Nut	72950-17
27	Chain & Stopper S/A	8174-02

● NOT AVAILABLE

Note - For Pop-up Drain parts refer to 2420.016 (N 2542)
For Chain & Stopper Drain parts refer to 2440.014 (N 2522)

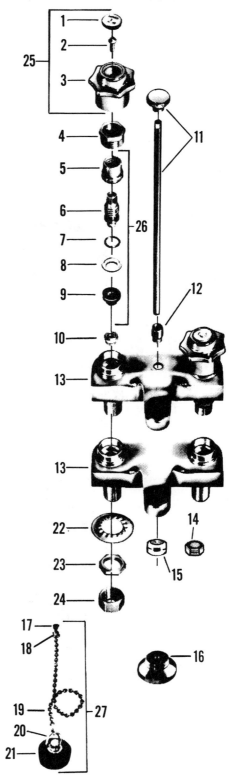

Fig. 2-14. American Standard centerset lavatory faucet.

Courtesy American Standard

Single Control Lavatory Fitting

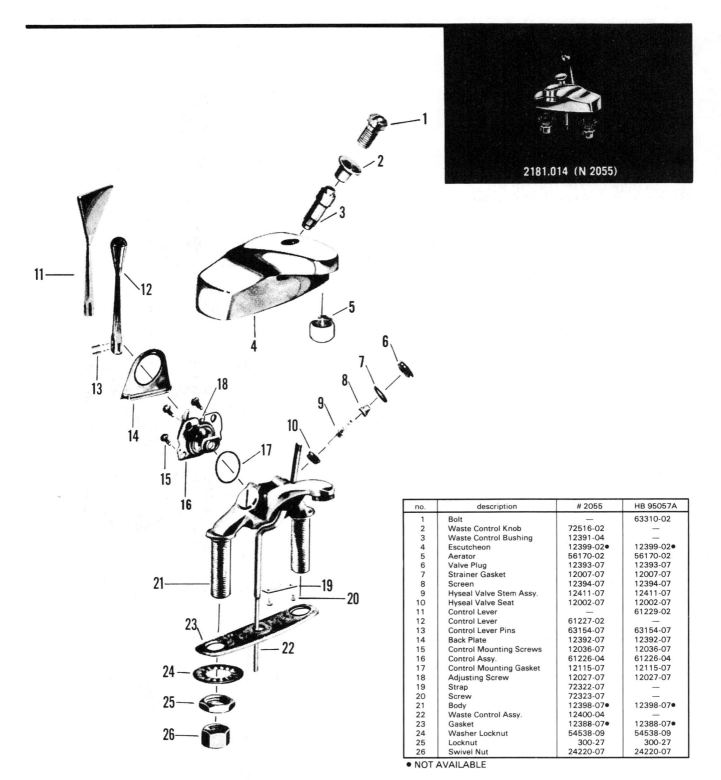

2181.014 (N 2055)

no.	description	# 2055	HB 95057A
1	Bolt	—	63310-02
2	Waste Control Knob	72516-02	—
3	Waste Control Bushing	12391-04	—
4	Escutcheon	12399-02●	12399-02●
5	Aerator	56170-02	56170-02
6	Valve Plug	12393-07	12393-07
7	Strainer Gasket	12007-07	12007-07
8	Screen	12394-07	12394-07
9	Hyseal Valve Stem Assy.	12411-07	12411-07
10	Hyseal Valve Seat	12002-07	12002-07
11	Control Lever	—	61229-02
12	Control Lever	61227-02	—
13	Control Lever Pins	63154-07	63154-07
14	Back Plate	12392-07	12392-07
15	Control Mounting Screws	12036-07	12036-07
16	Control Assy.	61226-04	61226-04
17	Control Mounting Gasket	12115-07	12115-07
18	Adjusting Screw	12027-07	12027-07
19	Strap	72322-07	—
20	Screw	72323-07	—
21	Body	12398-07●	12398-07●
22	Waste Control Assy.	12400-04	—
23	Gasket	12388-07●	12388-07●
24	Washer Locknut	54538-09	54538-09
25	Locknut	300-27	300-27
26	Swivel Nut	24220-07	24220-07

● NOT AVAILABLE

Note - Repair Kit consisting of ((2) each parts 7, 8, 9, 10, - #12417-07

Courtesy American Standard

Fig. 2-15. American Standard single-lever lavatory faucet.

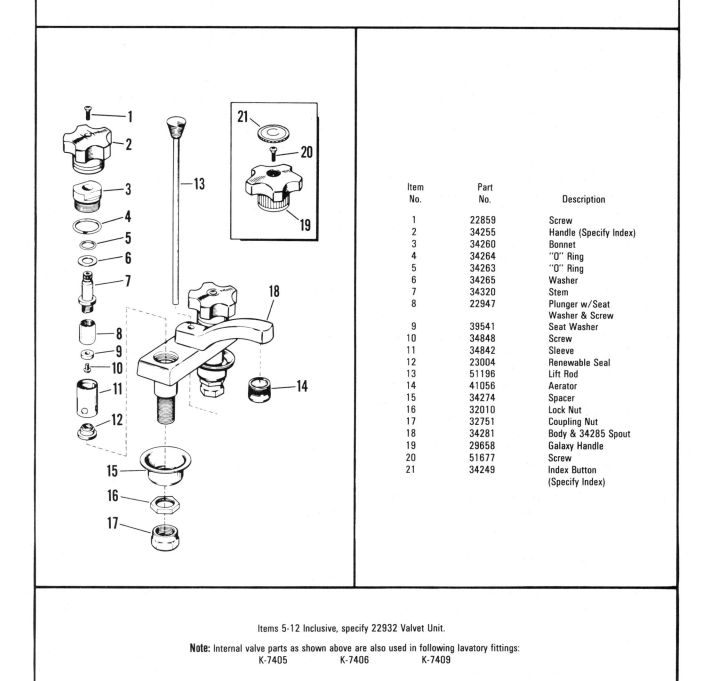

CONSTELLATION AND GALAXY CENTRA LAVATORY
FITTINGS PARTS
K-7400 & K-6960 FITTINGS

Item No.	Part No.	Description
1	22859	Screw
2	34255	Handle (Specify Index)
3	34260	Bonnet
4	34264	"O" Ring
5	34263	"O" Ring
6	34265	Washer
7	34320	Stem
8	22947	Plunger w/Seat Washer & Screw
9	39541	Seat Washer
10	34848	Screw
11	34842	Sleeve
12	23004	Renewable Seal
13	51196	Lift Rod
14	41056	Aerator
15	34274	Spacer
16	32010	Lock Nut
17	32751	Coupling Nut
18	34281	Body & 34285 Spout
19	29658	Galaxy Handle
20	51677	Screw
21	34249	Index Button (Specify Index)

Items 5-12 Inclusive, specify 22932 Valvet Unit.

Note: Internal valve parts as shown above are also used in following lavatory fittings:
K-7405   K-7406   K-7409

Fig. 2-16. Kohler centerset lavatory faucet.

Part Nos. 34264 and 34263, should also be replaced when installing new seat washers. When reassembling the faucet, note that the sleeve is grooved to receive the plunger. Lightly coat the new O-rings with waterproof grease to prevent damage when reassembling the faucet.

Kohler Triton Bancroft Lavatory Faucets
(K-7436, K-7437, K-7439)

PROBLEM: Faucet drips (Fig. 2-17).
CURE: Replace the worn seat washers, item 11; inspect the renewable seats if they are chipped or pitted.

Remove the screw (1) and the handle. Use an adjustable wrench to turn the bonnet (item 5) counterclockwise to loosen and remove. When the stem (item 9) is screwed into the plunger (item 10), the plunger can be lifted out of the faucet body. Lift the sleeve (item 13) out and inspect the washer seat. The washer seat should be replaced with a new seat, Part No. 23004, if it is chipped or pitted. The seat is pressed into the bottom of the sleeve. To remove the worn seat, insert a nail set through two opposite holes in the bottom of the sleeve and tap the square head of the nail set lightly with a small hammer. The worn seat will be forced out of the sleeve. Insert the new seat into the sleeve and tap it lightly to "set" it. Replace the worn seat washer (item 11). The two O-rings (items 6 and 7), Parts No. 34300 and 34263, should also be replaced when installing new seat washers. When reassembling the faucet, note that the sleeve is grooved to receive the plunger. The stems and plungers on the hot-water side of the faucet are different from those on the cold side; therefore, it is best to repair and reassemble one side of the faucet before repairing the other side to avoid mixing the parts. Lightly coat the new O-rings with waterproof grease before reassembly.

Kohler Triton Shelf Back Lavatory Faucets
K-8040

PROBLEM: Faucet drips (Fig. 2-18).
CURE: Replace the seat washers, inspect the renewable seats. If the seats are chipped or pitted, replace the seats.

Remove the screw (1) and the handle. Use an adjustable wrench to turn the bonnet (3) counterclockwise to loosen and remove. When the stem (7) is screwed into the plunger (8), the stem and the plunger can be pulled out of the faucet body. Pull the sleeve (11) out and inspect the renewable seat. If the seat is chipped or pitted, replace the seat with Part No. 23004. The seat is pressed into the bottom of the sleeve. To remove the worn seat, insert a nail set through two holes in the bottom of the sleeve and tap the square head of the nail set lightly with a small hammer. The worn seat will be forced out. Insert the new seat into the sleeve and tap it lightly to "set" it. It is advisable to replace the O-rings (4 and 5), Part Nos. 34300 and 34263, when installing a new seat washer. Lightly coat the new O-rings with waterproof grease before reassembling the faucet. Note that the sleeve is grooved to receive the plunger when reassembling. Because the hot and cold stems are different, it is best to repair and reassemble one side of the faucet before repairing the other side.

Bathtub Faucets (overrim fillers)
(bearing no brand name)

Fig. 2-19 shows a side view of a common type of overrim filler and an exploded view of the same faucet. Repair of this type of faucet is usually very easy; the main problem is getting it apart. It becomes even more of a problem if the bathroom has been remodeled and the walls have been tiled. This causes the faucet to be recessed into the wall the thickness of the tile. Very often when this has been done, the bonnet nut is behind the wall or tile and is very hard to reach with a regular wrench. First, remove the handle and escutcheon. If there is no nut holding the escutcheon on, the nipple is either made in one piece with the escutcheon, or it screws into the escutcheon from the back side. In this case, turn the escutcheon counterclockwise to unscrew it from the packing gland nut. Note: It is not necessary to remove the packing gland nut in order to remove the stem.

If the bonnet nut is accessible from the front of the faucet, use an adjustable wrench and turn the nut counterclockwise to unscrew it. If the bonnet nut is behind the wall, try to reach it through the access door (if there is one) in the wall behind the tub. If this is impossible, carefully enlarge the hole in the wall around the stem and under the escutcheon and carefully work the basin wrench, shown in Fig. 1-16, through the hole until you can lock the jaws of the basin wrench around the bonnet nut.

Turn the wrench counterclockwise to remove the nut. Turn the stem in the same direction as opening the faucet to remove the stem and install a new bibb washer. Inspect the seat; if it needs replacing or resurfacing, proceed as outlined in this chapter on **How to Reseat a Faucet.**

It is very unlikely that the faucet stems will require any new packing. If there is a water leak around the stem, tighten the packing gland by turning it ¼ turn clockwise. If you wish to add more packing, turn the packing gland counterclockwise to loosen and remove the gland. Wind a strand of graphited packing approximately 1½ in. long around the stem of the faucet, slide the packing gland back onto the faucet stem, and

TRITON BANCROFT LAVATORY FITTING PARTS WITH POP-UP DRAIN
K-7436, K-7437 & K-7439 Bancroft

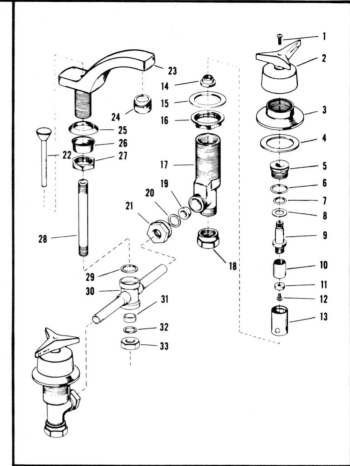

Item No.	Part No.	Description
1	41110	Screw (Round Head)
	53448	Screw (Countersunk)
2	41107	Triton Handle(Specify Index)
3	34304	Escutcheon
4	35715	Washer
5	20475	Bonnet
6	34300	"O" Ring
7	34263	"O" Ring
8	34265	Washer
9	34320	Hot Stem
	34319	Cold Stem
10	22947	Hot Plunger w/Seat Washer & Screw
	22948	Cold Plunger w/Seat Washer & Screw
11	39541	Seat Washer
12	34848	Screw
13	34842	Sleeve
14	23004	Renewable Seat
15	32060	Washer
16	33907	Lock Nut
17	34303	valve Body for K-7436 & K-7439— 4¼" Overall
	20480	Valve Body for K-7437—5" Overall
18	32751	Coupling Nut
19	20564	Slip Joint Washer
20	20565	Friction Ring
21	32751	Coupling Nut
22	51196	Lift Rod
23	35800	Spout w/Lift Rod Sleeve for K-7436 Spout Dimension 3⅝" c-c, Spout Shank 1¾"
	35795	Spout w/Lift Rod Sleeve for K-7437 Spout Dimension 4⅝" c-c, Spout Shank 2½"
	35798	Spout w/Lift Rod Sleeve for K-7439 Spout Dimension 4⅝" c-c, Spout Shank 1¾"
24	41007	Aerator
25	33504	Washer
26	34296	Spacer

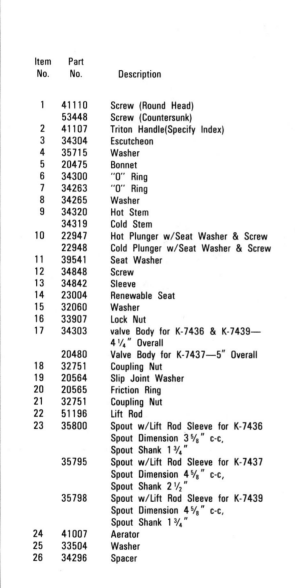

Item No.	Part No.	Description
27	33505	Lock Nut
28	34307	Lift Rod Sleeve for K-7436 & K-7439— 4" Long
	34971	Lift Rod Sleeve for K-7437— 4¾" Long
29	33031	Gasket
30	35821	Cross Connection Sub-Assembly
31	34294	Packing
32	34295	Friction Ring
33	34309	Lock Nut

Items 5-17, for K-7436 and K-7439, specify 34342 Hot Valve, 34343 Cold Valve.

Items 5-17, For K-7437, specify 20481 Hot Valve, 20684 Cold Valve.

Items 7-14 Inclusive, specify 22932 Valvet for Hot Valve, 22917 Valvet for Cold Valve.

Note: Internal valve parts as shown above are also used in following lavatory fittings:
K-7444   K-7445   K-7446   K-7447

Courtesy Kohler Co.

Fig. 2-17. Kohler adjustable spread lavatory faucet.

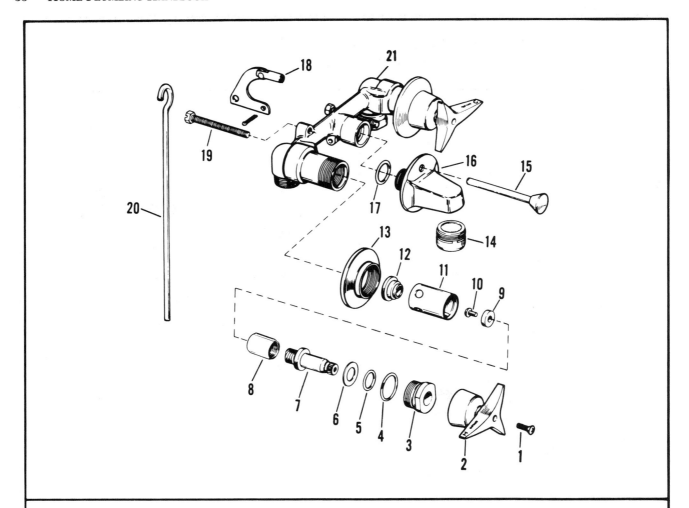

TRITON SHELF LAVATORY FITTING PARTS
For K-8040 Fitting
For Gramercy, Juneau, Strand, Rockport, Hampton, Taunton, Marston Lavatories

ITEM NO.	PART NO.	DESCRIPTION	ITEM NO.	PART NO.	DESCRIPTION
1	41110	Screw (Round Head)	10	34848	Screw
	53448	Screw (Countersunk)	11	34842	Sleeve
2	41105	Triton Handle (Specify Index)	12	23004	Renewable Seat
3	20475	Bonnet	13	34301	Flange
4	34300	"O" Ring	14	41007	Aerator
5	34263	"O" Ring	15	34349	Push Rod
6	34265	Washer	16	34435	Spout w/Aerator & "O" Ring
7	34320	Cold Stem	17	40933	"O" Ring
	34319	Hot Stem	18	34898	Trip Lever with #33799 Cotter Key
8	22947	Cold Plunger w/Seat Washer & Screw	19	35227	Screw
	22948	Hot Plunger w/Seat Washer & Screw	20	51198	Lift Rod
9	39541	Seat Washer	21	34298	Valve Body Only

Items 5-12 inclusive, specify 22932 Valvet for Cold Valve, 22917 Valvet for Hot Valve.

Fig. 2-18. Kohler shelf lavatory faucet.

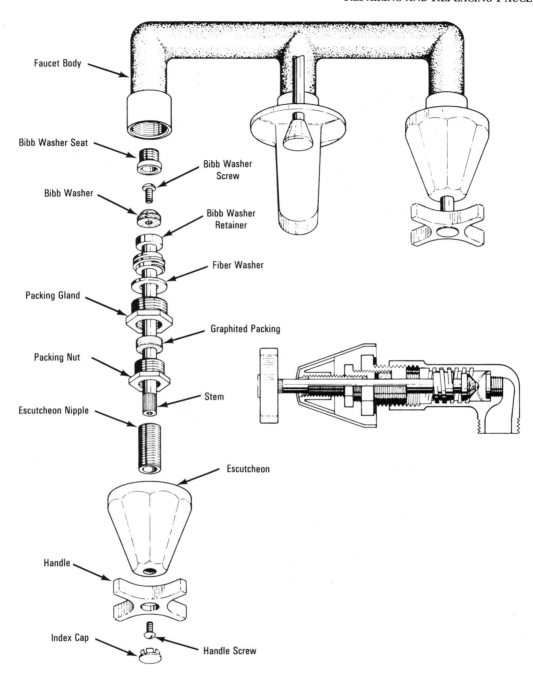

Fig. 2-19. Common type of bathtub faucet.

push the new strand of packing into the body of the faucet. The packing will be forced into place and shaped as you tighten the packing gland. Use a small adjustable wrench to turn the packing gland clockwise, and tighten the gland only until it is snug—do not overtighten. If the gland is too tight, the faucet stem will be hard to turn. Turn the hot and cold stems in the direction of opening the faucet until they are in the full open position when you reassemble the faucet. This will prevent damage to the stems when the bonnet nut is tightened. Reassemble the faucet, and the repair will be completed. If the diverter spout is defective, re-

place it; repair is impractical. Use a 14 in. pipe wrench and turn the old spout counterclockwise to loosen and remove the spout. Your local plumbing shop or hardware store should stock diverter spouts, and it is always a good idea when buying replacement parts to take the old parts with you; this insures that you will get the right replacement. Use pipe thread cement on the threads when you install the new spout. To avoid marring the chrome finish on the new spout, insert the handle of a long screwdriver into the spout end and turn the spout clockwise to tighten it. Since there is no water pressure on the water running out of the spout,

the spout should only be tightened until it touches the wall.

Kohler Shower and Bath Valve (overrim filler)
K-7032, K-7100, K-7210

PROBLEM: Faucets leak—drip (Fig. 2-20).
CURE: Replace the seat washer.

Remove the handle screw (1) and handle (2). Turn the escutcheon counterclockwise to loosen and remove. Use an adjustable wrench to turn the bonnet nut (7) counterclockwise to loosen and remove. Place the handle back on the stem and turn the stem in the direction of opening the faucet. The stem will come out of the faucet body. Replace the seat washer (12) with a new washer. If the seat (Remove Unit) is chipped or pitted, replace it with new part (item 14), Part No. 32462.

If the faucet leaks around the stem, it is usually only necessary to tighten the packing nut (5), turning it a half or a full turn clockwise to stop the leak. If the packing is badly worn, replace it with Part No. 32005. Reassemble the faucet.

Kohler Triton Diverter Shower and Bath Faucet
K-7014

PROBLEM: Faucet leaks (Fig. 2-21).
CURE: Replace the seat washers, inspect the seats. If the seats are chipped or pitted, replace the seats.

Remove the screw (1) and the handle (2). Use an adjustable wrench to turn the bonnet counterclockwise to loosen and remove it. When the plunger is screwed onto the stem, the stem and plunger can be pulled out together. Pull the sleeve (12) out and inspect the seat. If the seat is chipped or pitted, replace it with new part (13), Part No. 23004. Insert a nail set through two opposite holes in the sleeve, tap the square head of the nail set lightly with a small hammer, and the worn seat will be forced out of the sleeve. Insert the new seat into the sleeve and tap it lightly with a small hammer to "set" it. Replace the washer (10). Replace the O-rings (5 and 6), Part Nos. 34264 and 34263. Lightly coat the O-rings with waterproof grease to prevent damage when reassembling the faucet. It is unlikely that the diverter assembly will require repair. However, if this is necessary, refer to the illustration; start with the screw (14) and disassemble the diverter assembly. Replace the worn parts, if needed, and grease the O-rings (18 and 19) when reassembling the diverter.

HOW TO RESEAT A FAUCET

Refer to Fig. 2-22 showing reseating tool. At this point you have already disassembled the faucet and determined the need for resurfacing the bibb washer seat. The next step is to assemble the reseating tool to fit your faucet. Note that the female threads in the adapter collar (e) and (f) are different sizes and one of them will fit the bonnet threads of most faucets. If they do not fit, then the adapter (f) can be inverted and threaded onto the stem of the resurfacing tool and the male threads of the adapter can be threaded into the faucet body. Remove the old bibb washer from the washer retainer or the faucet stem and, using the bibb washer retainer as a guide, select the proper cutter. Thread the cutter on the stem of the reseating tool, turning it clockwise to tighten it. It is now assembled for use. Turn the stem of the reseating tool counterclockwise until the top of the cutting tool is resting against the bottom of the adapter. Insert the tool into the top of the faucet and, turning the adapter clockwise, tighten it hand-tight on the faucet body. Grasp the Tee handle on the top of the tool and turn the stem of the tool clockwise until you feel the cutter just touch the seat. Tighten the thumbscrew (g) on the collar (h) finger-tight. Then, with one hand on the Tee handle and the other grasping the collar (h), turn them both at the same time, clockwise, until you can no longer feel the cutter touching metal. Loosen the thumbscrew (g) and turn the Tee handle approximately $1/16$ of a turn clockwise; tighten the thumbscrew and again, as before, turn them both together clockwise until you can no longer feel the cutter touching metal. It is very important that both of these parts, the Tee handle and the collar, be turned together. This prevents the cutter from cutting too deeply into the metal of the bibb washer seat.

Remove the tool from the faucet body and, using the beam of a flashlight, inspect the washer seat. The seat should be bright, round, and polished. If any pits still show on the washer seat, repeat the operation.

When all pits, chips, etc., have been removed and the washer seat presents a bright, round, and unblemished surface, the seat repair is completed. Pour water into the top of the faucet to flush out all the metallic cuttings. It is very important that all metal cuttings be flushed out of the faucet; if these cuttings are left in the faucet, they will embed themselves in the soft rubber bibb washer and the faucet will leak. When the faucet has been thoroughly flushed out, install a new bibb washer in the bibb washer retainer on the faucet stem, and reassemble the faucet. Repair is now complete.

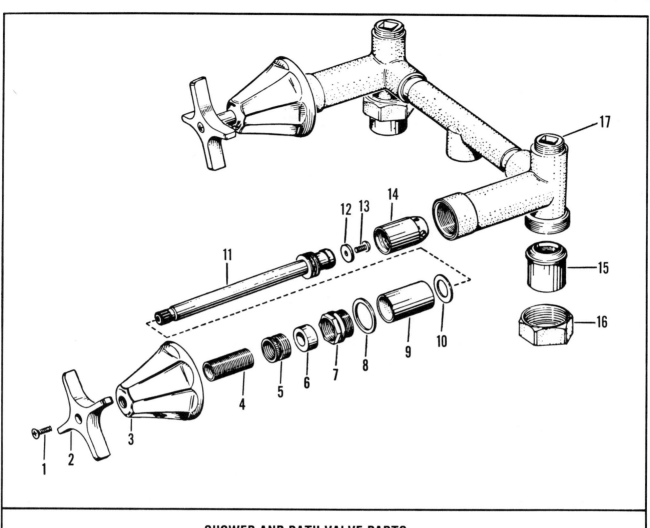

SHOWER AND BATH VALVE PARTS

For { K-7032 Dalton Shower and Bath Fitting
K-7100 Denton Bath Fitting
K-7210 Tipton Shower Fitting }

Item No.	Part No.	Description		Item No.	Part No.	Description
1	33357	Handle Screw		11	20242	Stem w/Washer & Screw
2	20419	Handle (specify index)		12	39541	Seat Washer
3	34116	Escutcheon		13	31490	Screw
4	22511	Adjusting Sleeve	Adjusting Sleeve & Packing Nut Assembly No. 20241	14	32462	Remov Unit
5	32008	Packing Nut		15	24015	Union Joint for ½″ I.P.
6	32005	Stem Packing			24899	Union Joint for ½″ O.D. Copper Tube
7	32002	Bonnet Nut				
8	32057	Gasket		16	24003	Nut
9	34190	Spacer		17	34191	Plug
10	32470	Washer (not required with new style #34190 spacer)			K-9800	Socket Wrench for #32002 Bonnet Nut (not shown)

Fig. 2-20. Kohler shower and bath faucet.

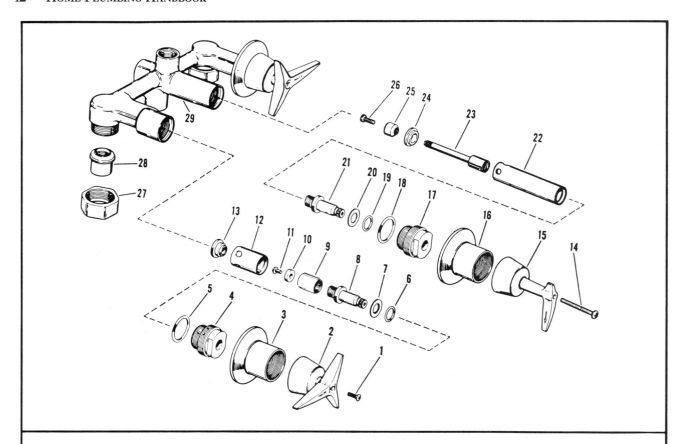

TRITON DIVERTER SHOWER AND BATH VALVE PARTS
K-7014 Dainey Shower and Bath Fitting

Item No.	Part No.	Description		Item No.	Part No.	Description
1	41110	Screw (Round Head)		14	41117	Screw
	53448	Screw (Countersunk)		15	41119	Diverter Handle
2	41099	Triton Handle (Specify Index)		16	22904	Escutcheon
*3	22904	Escutcheon		17	34253	Bonnet
4	34253	Bonnet		18	34264	"O" Ring
5	34264	"O" Ring		19	34263	"O" Ring
6	34263	"O" Ring		20	34265	Washer
7	34265	Washer		21	34320	Stem
8	34320	Cold Stem		22	22924	Sleeve with 23057 Sleeve
	34319	Hot Stem		23	22967	Stem
9	22947	Cold Plunger w/Seat		24	22927	Renewable Seat
		Washer & Screw		25	22942	Diverter Seat
10	22948	Hot Plunger w/Seat Washer & Screw		26	31490	Screw with 22941 Washer
11	34848	Screw		27	24003	Union Nut
12	34842	Sleeve		28	24015	Union for ½" I.P.
13	23004	Renewable Seat			24899	Union for ⅝" O.D. Copper Tube
				29	22928	Valve Body only

Items 6-13, specify 22932 Valvet for Cold Valve,
22917 Valvet for Hot Valve.

*Fitting with Screwdriver Stops, specify #29516 Escutcheon
and #29514 Sleeve.

Fig. 2-21. Kohler diverter (transfer valve) shower and bath faucet.

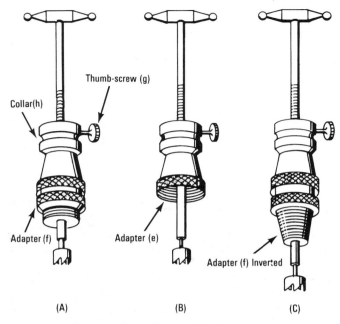

Thumb-screw (g)

Collar(h)

Adapter (f)

Adapter (e)

Adapter (f) Inverted

(A)   (B)   (C)

(A) Tool Assembled with Adapter (f) to Mount on Outside Male Threads of Small Bodied Faucets

(B) Tool Assembled without Adapter (f), Allowing Female Threads Inside Adapter (e) to Fit Male Threads on Larger Faucets

(C) Tool Assembled with Adapter (f) Inverted to Thread into Body of Faucet

Fig. 2-22. Reseating tool.

AERATORS

Aerators on faucet spouts serve a dual purpose. Aerators direct the flow of water from the spout to eliminate splashing, and they screen the flow of water to trap sand or other foreign substances.

Faucet spouts that are equipped with aerators may have either internal or external threads on the aerator end of the spout. An aerator can be added to a faucet spout that is not threaded. A universal "clip-adapter-type" aerator can be used.

If the nylon or copper screens in an aerator become obstructed, the water flow through the aerator will be decreased. Remove the aerator body from the spout. Using slip joint pliers to grasp the body of the aerator firmly, turn the aerator clockwise (viewed from the front and above) to loosen and remove the aerator body. Remove the nylon and copper screens and wash out the foreign material caught in the screens. When removing the screens, note the order in which they are assembled in the aerator body. The screens and washers must be reinstalled in the aerator in the same order

as they were removed, for the aerator to work properly.

REMOVING OLD SINK AND LAVATORY FAUCETS

Faucets designed for the do-it-yourself trade are sold in the plumbing departments of hardware and building supply stores. Step-by-step installation instructions are provided. The satisfaction of being able to do it yourself, plus the dollars saved, make the installation of new faucets a worthwhile project.

The removal of the old faucet may be much more difficult than the installation of the new one. The following instructions will simplify the removal process. First, shut off the hot- and cold-water supply to the old faucet and disconnect the water supply lines from the faucet. Bottom-mounted faucets are secured to the sink by locknuts, as shown in Fig. 2-23. The locknuts may be corroded and rusted. Penetrating oil sprayed on the locknuts and shanks will aid in loosening and removing the nuts. A basin wrench should be used to turn the nuts and the nuts should be turned counterclockwise (viewed from below) to loosen and remove them. When the locknuts are removed, the old faucet can be lifted up and off of the sink.

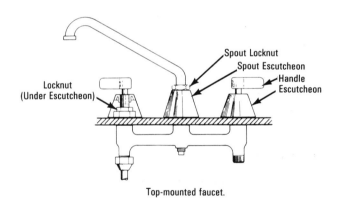

Locknut
(Under Escutcheon)

Spout Locknut
Spout Escutcheon
Handle
Escutcheon

Top-mounted faucet.

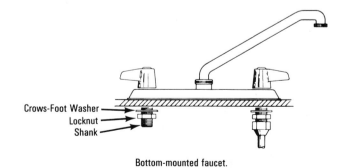

Crows-Foot Washer
Locknut
Shank

Bottom-mounted faucet.

Fig. 2-23. Top- and bottom-mount sink faucets.

A top-mounted faucet is shown in Fig. 2-23. The faucet is secured to the sink by a lockwasher located under each handle escutcheon. To remove this type of faucet, turn off and disconnect the water supplies to the faucet. Loosen and remove the spout locknut, lift the spout out, and loosen and remove the spout escutcheon. Remove the handles and turn the handle escutcheons counterclockwise to loosen and remove them. Turn the locknuts under the escutcheons counterclockwise to loosen and remove them. The faucet can then be removed from under the sink.

HINTS ON INSTALLING OR REPAIRING FAUCETS

Information on the faucets shown on the following pages is presented to help the reader identify the faucet he or she is now using or to help in the selection and installation of a new or replacement faucet. Delta, Delex, Peerless, and Moen faucets are being used extensively in new homes and to replace older types of faucets in existing homes. At some time in its life almost every faucet needs repair; knowing how to disassemble a faucet, what parts are needed, and how to install these parts and reassemble the faucet are the keys to a successful repair job. The faucets shown on the following pages are all "washerless" types; parts are available at plumbing shops and building supply stores, and the installation and repair information packaged with each faucet, or found in the following pages, will simplify the work.

The following "helpful hints," acquired over years of experience, not always found in the instructions for installation and repair furnished with new faucets or repair parts, will make the job easier.

Kitchen Sink Faucets

If you have a kitchen sink with four mounting holes and want to install a new faucet that does not have a spray hose, a chrome-plated device (called a "cock hole cover"), available at plumbing shops and building supply stores, will cover the fourth or spray-hose hole. When a kitchen faucet with a spray hose is used, the spray head rests in a spray-hose support mounted in the fourth hole in the sink top. Before the hose support is inserted through the sink top, the base or cup of the support should be filled with soft plumber's putty. When the support is inserted through the sink top and tightened, the putty will prevent water leakage through the support hole.

When connecting a spray hose to a faucet, the job will be much easier if the faucet is held above the sink top, the spray hose inserted down through the support, and then the end of the faucet. The hose can then be easily connected to the faucet, and the faucet then set into place and secured to the sink.

When a faucet that uses a cap with an adjusting ring to stop leakage around the valve (Delta, Peerless—Fig. 2-24) is first installed or is disassembled for repairs, the cap and the adjusting ring should be removed and both female threads in the cap should be lightly coated with a light grease. This will help prevent corrosion

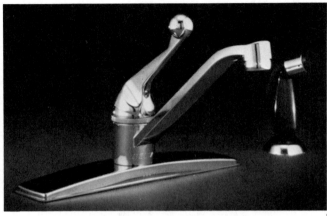

Fig. 2-24. Delta single-lever sink faucet with spray head.

forming between the adjusting ring and the cap and between the cap and the faucet body threads, and will make initial adjustment and future repairs much easier. Also, on these faucets, when the cap is replaced it should be tightened only *hand tight*. Any water leakage at this point is controlled by tightening the adjusting ring. A special wrench with two prongs on one end and an Allen wrench on the other end is furnished with each new Delta or Peerless single-hand or single-lever faucet. This wrench will be needed to remove the handle or lever and to turn the adjusting ring in the cap of these faucets. If you do not have this wrench, it is available at stores selling Delta or Peerless products at little or no charge. On some older-model faucets, the adjusting ring may be made of brass, and if the threads in the cap were not greased, as mentioned earlier, it may be impossible to turn the adjusting ring. If you encounter this problem, a new cap will probably be needed.

When any new faucet is installed or an older one is repaired, the water supplies to the faucet must be turned off before work is started. After either a new installation or repair work on the faucet, *before the water supplies to the faucet are turned on*, the aerator on the end of the spout should be removed. Be careful

DELTA®

INSTALLATION GUIDE

SINGLE
LEVER
KITCHEN FAUCETS

MODELS: 100, 110, 300,
400, 100 HDF

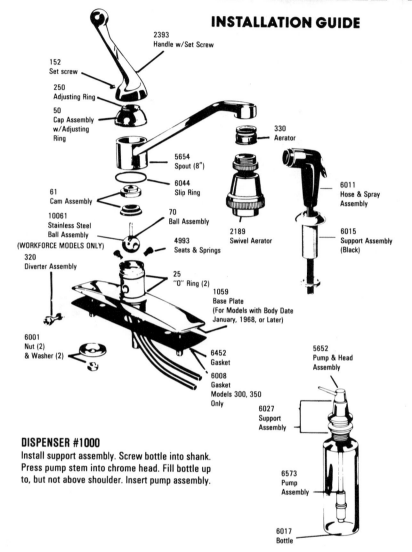

2393
Handle w/Set Screw

152
Set screw

250
Adjusting Ring

50
Cap Assembly
w/Adjusting
Ring

61
Cam Assembly

10061
Stainless Steel
Ball Assembly
(WORKFORCE MODELS ONLY)

320
Diverter Assembly

6001
Nut (2)
& Washer (2)

330
Aerator

5654
Spout (8")

6044
Slip Ring

70
Ball Assembly

4993
Seats & Springs

2189
Swivel Aerator

25
"O" Ring (2)

1059
Base Plate
(For Models with Body Date
January, 1968, or Later)

6452
Gasket

6008
Gasket
Models 300, 350
Only

6011
Hose & Spray
Assembly

6015
Support Assembly
(Black)

INSTALLATION STEPS

1. Install spray support into hole of sink.

2. Before installing faucet, insert spray hose through support and up through center hole of sink. Connect hose to nipple on faucet.

3. **CAUTION:** Do not straighten inlet tubes before installing faucet. Feed inlet tubes down through sink deck holes. Secure with washers and nuts. Then, with EXTREME care, bend copper tubing.

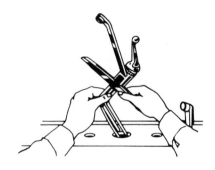

6" CENTER ADJUSTMENT—Models 100, 400
Faucets are furnished for 8" center decks. To adjust for 6 " center, remove the screws from base plate and move bolts to inside 6" center holes.

5652
Pump & Head
Assembly

6027
Support
Assembly

6573
Pump
Assembly

6017
Bottle

DISPENSER #1000
Install support assembly. Screw bottle into shank. Press pump stem into chrome head. Fill bottle up to, but not above shoulder. Insert pump assembly.

Courtesy Delta Faucet Co.

Routine Maintenance Instructions

Delta Faucets have earned the reputation of superiority in design, engineering, performance and durability in millions of home installations. Routine faucet maintenance to compensate for foreign materials in varying water conditions, will restore like new performance and extend the life of the faucet. Ease and simplicity of service and maintenance is another advantage of having Delta faucets throughout the home.

A. If you should have a leak under handle—tighten adjusting ring by following steps 1, 14, 15 and 16.

B. If you should have a leak from a spout—replace seats and springs by following steps 1 2, 3, 4 and 10. Then reassemble with steps 11, 12, 13, 14, 15 and 16.

C. If you should have a leak from around spout collar—replace "O" rings by following steps 1, 2, 5, 6 and 9. Then reassemble by following steps 13, 14, 15 and 16.

D. If spray does not work properly—check hose for wear or cuts. Clean or replace diverter part No. 320 by following steps 1, 2, 5, 7, 8 and 9A. Then reassemble, following steps 13, 14, 15 and 16.

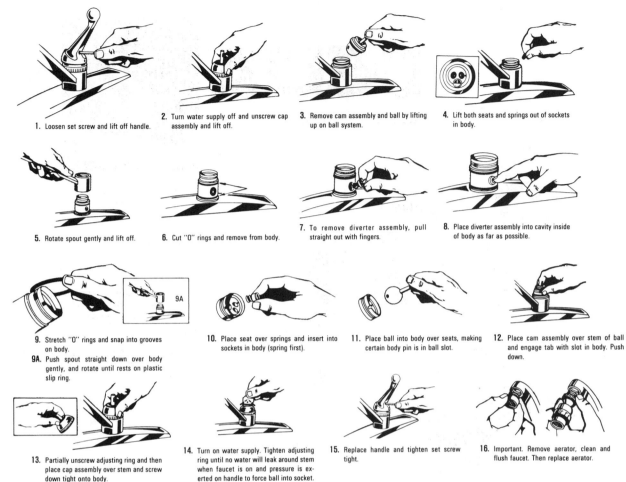

1. Loosen set screw and lift off handle.

2. Turn water supply off and unscrew cap assembly and lift off.

3. Remove cam assembly and ball by lifting up on ball system.

4. Lift both seats and springs out of sockets in body.

5. Rotate spout gently and lift off.

6. Cut "O" rings and remove from body.

7. To remove diverter assembly, pull straight out with fingers.

8. Place diverter assembly into cavity inside of body as far as possible.

9. Stretch "O" rings and snap into grooves on body.
9A. Push spout straight down over body gently, and rotate until rests on plastic slip ring.

10. Place seat over springs and insert into sockets in body (spring first).

11. Place ball into body over seats, making certain body pin is in ball slot.

12. Place cam assembly over stem of ball and engage tab with slot in body. Push down.

13. Partially unscrew adjusting ring and then place cap assembly over stem and screw down tight onto body.

14. Turn on water supply. Tighten adjusting ring until no water will leak around stem when faucet is on and pressure is exerted on handle to force ball into socket.

15. Replace handle and tighten set screw tight.

16. Important. Remove aerator, clean and flush faucet. Then replace aerator.

Courtesy Delta Faucet Co.

Routine maintenance/repair instructions for Delta single-lever sink faucet.

to keep the various parts in the aerator in place. If these parts are removed from the aerator and not re assembled in exactly the right order, the aerator will not work properly. With the aerator turned off, the water supplies should be turned on and the faucet should be flushed out for one minute. This will prevent any scale or debris loosened during the installation or repair process from stopping up or clogging the aerator. After flushing out the faucet, the aerator should be replaced.

Lavatory Faucets

When a lavatory faucet and pop-up waste is installed, a ring of soft putty should be placed under the pop-up flange, as shown in Fig. 14-3. One layer of Teflon tape dope should be wrapped around the threads on the pop-up body to which the pop-up flange connects, as shown in Fig. 2-28 (p. 59), before the pop-up waste is assembled. Various methods of

connecting the water supplies to a new lavatory are shown in Fig. 14-2.

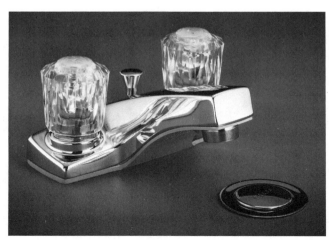

Courtesy Delta Faucet Company

Fig. 2-25. Delex two-handle lavatory faucet.

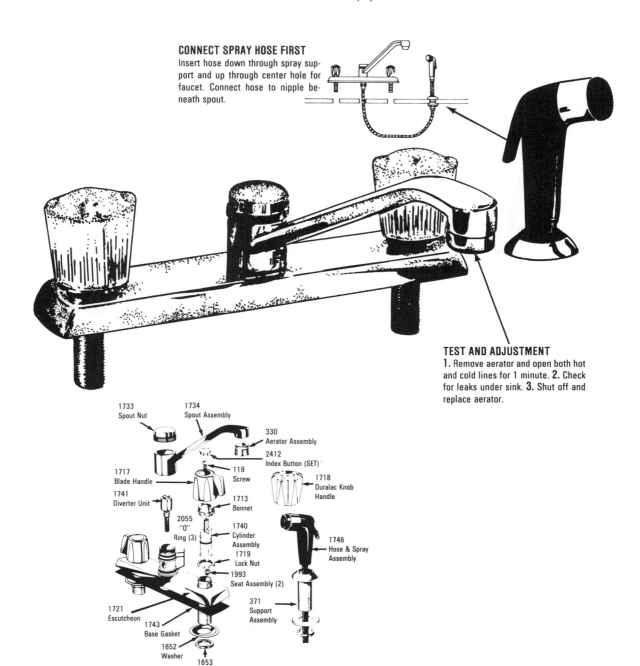

Installation Guide
with or without spray

CONNECT SPRAY HOSE FIRST
Insert hose down through spray support and up through center hole for faucet. Connect hose to nipple beneath spout.

TEST AND ADJUSTMENT
1. Remove aerator and open both hot and cold lines for 1 minute. **2.** Check for leaks under sink. **3.** Shut off and replace aerator.

1733 Spout Nut
1734 Spout Assembly
330 Aerator Assembly
2412 Index Button (SET)
119 Screw
1718 Duralac Knob Handle
1717 Blade Handle
1713 Bonnet
1741 Diverter Unit
2055 "O" Ring (3)
1740 Cylinder Assembly
1746 Hose & Spray Assembly
1719 Lock Nut
1993 Seat Assembly (2)
371 Support Assembly
1721 Escutcheon
1743 Base Gasket
1652 Washer
1653 Nut

Courtesy Delta Faucet Co.

Installation guide for Delex two-handle sink faucet.

Routine Maintenance Instructions

Delex faucets have earned the reputation of superiority in design, engineering, performance and durability in millions of home installations. Routine faucet maintenance, to compensate for foreign materials in varying water conditions, will restore like new performance and extend the life of the faucet. Ease and simplicity of service and maintenance is another advantage of having Delex faucets throughout the home.

A. If faucet leaks from under handle or from spout outlet, replace stem unit and/or seat as in steps 1, 2, 3, 4, 5. Reassemble following steps 12, 13 and 16.

B. If leak occurs at top or bottom of spout body, replace "O" Rings as shown in steps 6, 7, 8, 9, 14 and 15.

C. If volume of water from spray models decreases or spray ceases to function, follow procedure in steps 6, 10, 11 and 15.

SHUT OFF WATER SUPPLY

1. Pry off index button, remove screw and lift off handle.
2. Unscrew bonnet.
3. Pull stem straight up and out.
4. Lift seat and spring out of body.
5. Place new seat over new spring and push into socket in body.
6. Unscrew spout nut.
7. Lift spout off by rotating back and forth, and pull up gently.
8. Cut "O" rings and remove from body.
9. Stretch new "O" rings and snap into grooves on body.
10. Unscrew diverter using screwdriver. Wash off diverter, and clean and flush out socket of body.
11. Screw diverter tightly in spout socket.

12. IMPORTANT—STEM MUST BE REPLACED PROPERLY!

12A. **Knob Handle Stem Position**—Slip stem unit into body, aligning key with key way, so "stop" on both stems point toward spout.

12B. **Blade Handle Stem Position**—As you face faucet, slip stem units into body, aligning keys in key way so "stops" on both units point to right.

13. Replace bonnet. Tighten securely.

14. Push spout straight down over body, rotating back and forth gently until it rests on faucet base.
15. Screw spout nut on snugly.
16. Replace handle. Tighten screw. Press index button in position.

Use only genuine Delta replacement parts.
Kit 1740—Cylinder Assembly
Kit 1933—Seat Assembly (2)
Kit 2055—"O" Ring (3)

Courtesy Delta Faucet Co.

Routine maintenance/repair instructions for Delta two-handle sink faucet.

Test and Adjustment Procedure for Faucets Using Cap and Adjusting Ring

After the faucet has been installed or repaired and the supply piping connected, it should be tested and adjusted as follows:

1. Remove the aerator on the spout (bath and shower valves have no aerator) and open faucet to mixed position (knob, lever, or handle up and in center)
2. Turn on water for one minute.

DELTA®

INSTALLATION GUIDE

SINGLE
HANDLE
LAVATORY
CENTERSET
FAUCETS

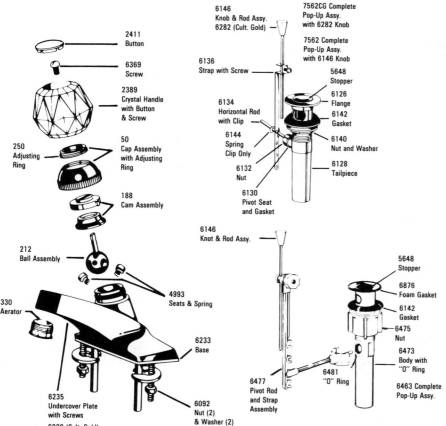

2411 Button

6369 Screw

2389 Crystal Handle with Button & Screw

250 Adjusting Ring

50 Cap Assembly with Adjusting Ring

188 Cam Assembly

212 Ball Assembly

330 Aerator

6235 Undercover Plate with Screws
6238 (Cult. Gold)

6146 Knob & Rod Assy.
6282 (Cult. Gold)

6136 Strap with Screw

6134 Horizontal Rod with Clip

6144 Spring Clip Only

6132 Nut

6130 Pivot Seat and Gasket

7562CG Complete Pop-Up Assy. with 6282 Knob

7562 Complete Pop-Up Assy. with 6146 Knob

5648 Stopper

6126 Flange

6142 Gasket

6140 Nut and Washer

6128 Tailpiece

4993 Seats & Spring

6233 Base

6092 Nut (2) & Washer (2)

6146 Knot & Rod Assy.

6477 Pivot Rod and Strap Assembly

6481 "O" Ring

5648 Stopper

6876 Foam Gasket

6142 Gasket

6475 Nut

6473 Body with "O" Ring

6463 Complete Pop-Up Assy.

Models: 502, 503, 522, 532

TEST AND ADJUSTMENT

1. Remove aerator and open valve to mixed position.

2. Turn on water for 1 minute.

3. Check for leaks under sink.

4. Shut off and replace aerator.

5. Open faucet and press down on handle toward valve.

6. If water squirts under handle, remove and tighten adjusting ring.

Courtesy Delta Faucet Co.

Routine Maintenance Instructions

Delta faucets have earned the reputation of superiority in design, engineering, performance and durability in millions of home installations. Routine faucet maintenance to compensate for foreign materials in varying water conditions, will restore like new performance and extend the life of the faucet. Ease and simplicity of service and maintenance is another advantage of having Delta faucets throughout the home.

A. If you should have a leak under handle—tighten adjusting ring following steps 1 and 9. Reassemble as in step 10.

B. If you should have a leak from spout—shut off water supply, and follow steps 1, 2, 3, 4 and 5. Reassemble as in steps 6, 7 and 8. Set adjusting ring as in 9. Replace handle as in 10.

1. Pry off handle button, remove screw and lift off handle.

2. Unscrew cap assembly and lift off.

3. Remove cam assembly and ball by lifting up on ball stem.

4. Lift both seats and springs out of sockets in body.

5. Place seat over springs and insert into sockets in body (spring first).

6. Place ball into body over seats.

7. Place cam assembly over stem of ball and engage tab with slot in body. Push down.

8. Partially unscrew adjusting ring and then place cap assembly over ball stem and screw down tight onto body.

9. Tighten adjusting ring until no water will leak around stem when faucet is on and pressure is exerted on handle to force ball into socket.

10. Replace handle. Tighten handle screw—tight. Replace handle button with "ON" arrow pointing to rear of lavatory faucet and pointing up on shower or tub and shower faucet.

Courtesy Delta Faucet Co.

Repair instructions for Delta single-handle lavatory faucet.

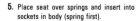

Delex BY DELTA
TWO HANDLE LAVATORY
Series 2500,

Installation Guide

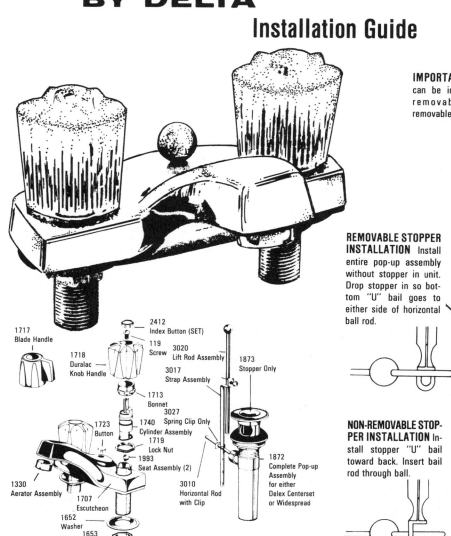

IMPORTANT—Stopper can be installed either removable or non-removable.

IMPORTANT—Lay ring of putty beneath pop-up flange before installing!

REMOVABLE STOPPER INSTALLATION Install entire pop-up assembly without stopper in unit. Drop stopper in so bottom "U" bail goes to either side of horizontal ball rod.

NON-REMOVABLE STOPPER INSTALLATION Install stopper "U" bail toward back. Insert bail rod through ball.

#3017 STRAP MAY BE BENT if necessary to connect properly with #1310 horizontal ball lift rod.

1717 Blade Handle

1718 Duralac Knob Handle

1723 Button

1330 Aerator Assembly

1707 Escutcheon

1652 Washer

1653 Nut

2412 Index Button (SET)

119 Screw

3020 Lift Rod Assembly

3017 Strap Assembly

1713 Bonnet

3027 Spring Clip Only

1740 Cylinder Assembly

1719 Lock Nut

1993 Seat Assembly (2)

3010 Horizontal Rod with Clip

1873 Stopper Only

1872 Complete Pop-up Assembly for either Delex Centerset or Widespread

Courtesy Delta Faucet Co.

Routine Maintenance Instructions

Delex faucets have earned the reputation of superiority in design, engineering, performance and durability in millions of home installations. Routine faucet maintenance, to compensate for foreign materials in varying water conditions, will restore like new performance and extend the life of the faucet. Ease and simplicity of service and maintenance is another advantage of having Delex faucets throughout the home.

A. If faucet leaks from under handle or from spout outlet, replace stem unit and/or seat as in steps 1, 2, 3, 4, 5. Reassemble following steps 6, 7 and 8.

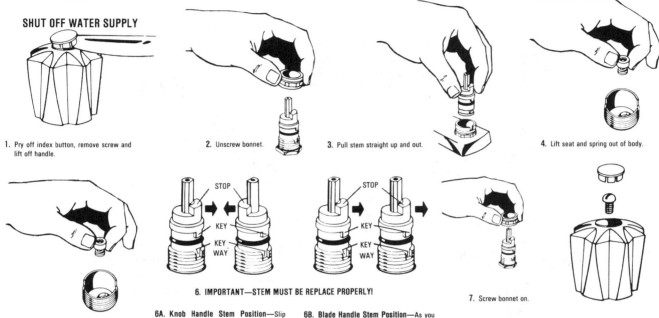

SHUT OFF WATER SUPPLY

1. Pry off index button, remove screw and lift off handle.

2. Unscrew bonnet.

3. Pull stem straight up and out.

4. Lift seat and spring out of body.

5. Place new seat over new spring and push into socket in body.

6. IMPORTANT—STEM MUST BE REPLACE PROPERLY!

6A. Knob Handle Stem Position—Slip stem unit into body, aligning key with key way, so "stop" on both stems point toward spout.

6B. Blade Handle Stem Position—As you face faucet, slip stem units into body, aligning keys in key way so "stops" on both units point to right.

7. Screw bonnet on.

8. Replace handle. Tighten screw. Press index button in position.

Courtesy Delta Faucet Co.

Repair instructions for Delta two-handle lavatory faucet.

3. Check for leaks under sink or lavatory.
4. Turn faucet off; knob, lever, or handle down and in center.
5. Open faucet and *press down* on knob, lever, or handle.
6. If water squirts from under knob, lever, or handle, remove the knob, lever, or handle and tighten the adjusting ring in the cap. The adjusting ring and cap is shown in both Delta and Peerless installation and routine maintenance precedures as illustrated in this chapter.

A Delex two-handle lavatory faucet is shown in Fig. 2-25 (p. 46).

A Delta single-handle bath valve is shown in Fig. 2-26.

A Moen single-lever kitchen faucet is shown in Fig. 2-27 (p. 58).

Installation and repair procedures for these faucets are also shown in this chapter.

Courtesy Delta Faucet Company

Fig. 2-26. Delta single-handle bath valve.

Delta SINGLE HANDLE BATH VALVES Series

Single handle with or without pressure balance

Installation Guide
(Pressure Balance Model Illustrated)

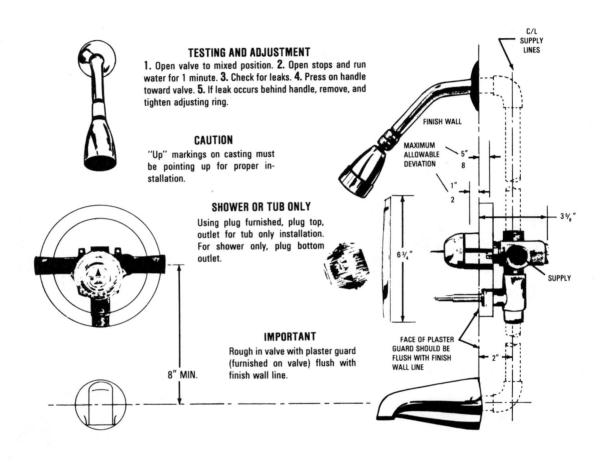

TESTING AND ADJUSTMENT
1. Open valve to mixed position. 2. Open stops and run water for 1 minute. 3. Check for leaks. 4. Press on handle toward valve. 5. If leak occurs behind handle, remove, and tighten adjusting ring.

CAUTION
"Up" markings on casting must be pointing up for proper installation.

SHOWER OR TUB ONLY
Using plug furnished, plug top, outlet for tub only installation. For shower only, plug bottom outlet.

IMPORTANT
Rough in valve with plaster guard (furnished on valve) flush with finish wall line.

C/L SUPPLY LINES

FINISH WALL

MAXIMUM ALLOWABLE DEVIATION   5" / 8

1" / 2

6¾"

3⅝"

SUPPLY

FACE OF PLASTER GUARD SHOULD BE FLUSH WITH FINISH WALL LINE

2"

8" MIN.

Delta Single Handle
Models 602, 606, 612, 622, 632, 636

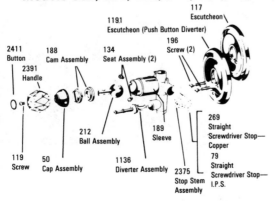

2411 Button
2391 Handle
188 Cam Assembly
134 Seat Assembly (2)
1191 Escutcheon (Push Button Diverter)
196 Screw (2)
117 Escutcheon
212 Ball Assembly
189 Sleeve
119 Screw
50 Cap Assembly
1136 Diverter Assembly
2375 Stop Stem Assembly
269 Straight Screwdriver Stop—Copper
79 Straight Screwdriver Stop—I.P.S.

Delta Single Handle Pressure Balance
Models 604, 608, 624, 634, 638

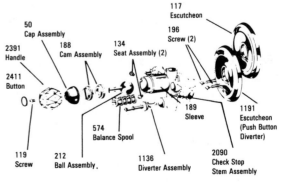

50 Cap Assembly
2391 Handle
2411 Button
188 Cam Assembly
134 Seat Assembly (2)
196 Screw (2)
117 Escutcheon
574 Balance Spool
189 Sleeve
1191 Escutcheon (Push Button Diverter)
119 Screw
212 Ball Assembly
1136 Diverter Assembly
2090 Check Stop Stem Assembly

Courtesy Delta Faucet Co.

Routine Maintenance Instructions

Delta faucets have earned the reputation of superiority in design, engineering, performance and durability in millions of home installations. Routine faucet maintenance, to compensate for foreign materials in varying water conditions can restore like new and extend the life and performance of the faucet. And, ease and simplicity of service and maintenance is another advantage of having Delta faucets throughout the home.

A. If you should have leak under handle—tighten adjusting ring by following steps 1 and 9 leaving water supply and faucet turned on. Replace handle as in step 10.

B. If you should have leak from spout—replace seats and springs by following steps 1, 2, 3 and 4. Reassemble following steps 5, 6, 7 and 8. Turn water supply on and follow steps 9 and 10.

Shut off water supply

1. Pry off handle button, remove screw and lift off handle.

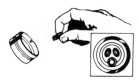

2. Unscrew cap assembly and lift off.

3. Remove cam assembly and ball by lifting up on ball stem.

4. Lift both seats and springs out of sockets in body.

To reassemble faucet

5. Place seat over springs and insert into sockets in body (spring first).

6. Place ball into body over seats.

7. Place cam assembly over stem of ball and engage tab with slot in body. Push down.

8. Partially unscrew adjusting ring and then place cap assembly over ball stem and screw down tight onto body.

Turn on water supply

9. Tighten ring until no water will leak around stem when faucet is on and pressure is exerted on handle to force ball into socket.

10. Replace handle. Tighten handle screw—tight. Replace handle button with "ON" arrow pointing up.

PRESSURE BALANCED VALVES

If unable to maintain constant temperature,
clean balancing spool as follows.

1. Remove handle and escutcheon.
2. Close stops.
3. Unscrew balancing spool assembly (part No. 574) and remove—slowly—so as not to damage "O" Ring Seals. NOTE. On shower only installations water tapped in shower riser will drain out when balancing spool assembly is removed.
4. Examine valve hole for chips or any other foreign matter.
5. Remove any chips from spool sleeve before attempting to remove spool from inside of sleeve.
6. Remove spool from sleeve and clean all de-posits from both sleeve and spool.
7. When spool will slide freely in sleeve, replace assembly in original position.
8. Open stops and check cap of spool assembly and stems of check stops for leaks before reinstal-ling escutcheon.

Courtesy Delta Faucet Co.

Repair instructions for Delta single-handle bath valve.

PEERLESS SINGLE HANDLE LAVATORY FAUCETS

If you should have a leak from the spout outlet—Replace the seats and springs, using Peerless repair kit #1815.

SHUT OFF WATER.

1. Remove handle, unscrew cap assembly and lift off.

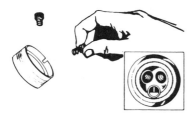

2. Lift out ball and cam assembly by pulling up on stem.

3. Remove old seats and springs and insert new ones. Reassemble faucet in inverse order.

If leak persists—Install a new ball, using Peerless repair kit #1963.

SHUT OFF WATER.

Cam

Packing

1. Remove handle and cap assembly and lift out old ball.

2. Slip packing and then cam on new ball and insert into socket. Reassemble faucet in reverse order.

A. If you should have a leak from under handle—

1. Do not shut off water. Remove handle and use Peerless wrench to tighten adjusting ring.

2. Tighten adjusting ring until no water will leak around cap assembly when faucet is on and pressure is exerted to force ball into socket.

Model 8620

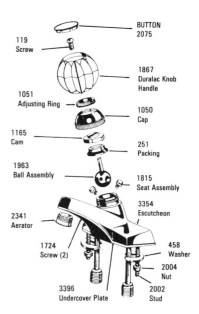

119
Screw

BUTTON
2075

1867
Duralac Knob
Handle

1051
Adjusting Ring

1050
Cap

1165
Cam

251
Packing

1963
Ball Assembly

1815
Seat Assembly

3354
Escutcheon

2341
Aerator

1724
Screw (2)

458
Washer

2004
Nut

3396
Undercover Plate

2002
Stud

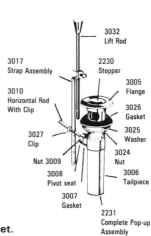

3032
Lift Rod

3017
Strap Assembly

2230
Stopper

3005
Flange

3010
Horizontal Rod
With Clip

3026
Gasket

3025
Washer

3027
Clip

3024
Nut

Nut 3009

3006
Tailpiece

3008
Pivot seat

3007
Gasket

2231
Complete Pop-up
Assembly

Repair instructions for Peerless single-handle lavatory faucet.

PEERLESS TWO HANDLE (DURALAC) LAVATORY FAUCETS

Model 9620

B. If you should have a leak from the spout outlet—replace the seats and springs using Peerless repair kit #1815.

1. Shut off water. Remove handle, unscrew cap assembly and lift off.

To Disassemble—shut off water.

1. Pry off handle button, remove screw and lift off handle.

Unscrew bonnet counter-clockwise.

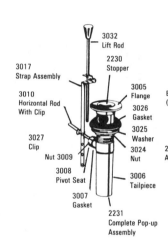

3032 Lift Rod
2230 Stopper
3017 Strap Assembly
3005 Flange
3010 Horizontal Rod With Clip
3026 Gasket
3025 Washer
3027 Clip
3024 Nut
Nut 3009
3008 Pivot Seat
3006 Tailpiece
3007 Gasket
2231 Complete Pop-up Assembly

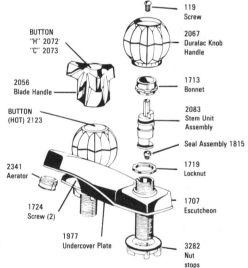

BUTTON "H" 2072 "C" 2073
2056 Blade Handle
BUTTON (HOT) 2123
BUTTON (COLD) 2124
119 Screw
2067 Duralac Knob Handle
1713 Bonnet
2083 Stem Unit Assembly
Seal Assembly 1815
1719 Locknut
2341 Aerator
1724 Screw (2)
1977 Undercover Plate
1707 Escutcheon
3282 Nut stops

If leak persists—Install new stem unit using Peerless repair kit #2083.

1. Remove handle and bonnet nut.

2. Lift out stem unit and replace with new stem unit.

Note: For **knob** handle faucets, both stops must point toward spout. For **blade** handle faucets, left hand stop points toward spout, right hand stop points away from spout.

3. Reassemble faucet in reverse order.

Stops

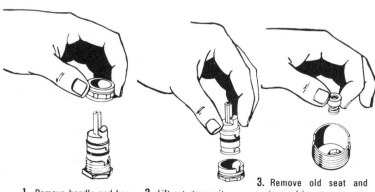

1. Remove handle and bonnet nut.

2. Lift out stem unit.

3. Remove old seat and spring and insert new ones. Reassemble faucet in reverse order.

Repair instructions for Peerless two-handle lavatory faucet.

Courtesy Peerless Faucets

PEERLESS SINGLE HANDLE KITCHEN FAUCETS

To Open Faucet For Service

Shut off
water supply.

1. Loosen set screw with Allen wrench end of Peerless wrench and lift off handle.

Unscrew cap assembly and lift off.

2. Lift out ball and cam assembly by pulling up on stem.

3. Remove old seats and springs and insert new ones. Reassemble faucet in reverse order.

If leak persists—install a new ball, using Peerless repair kit #1963.

1. Remove handle and cap assembly and lift out old ball.

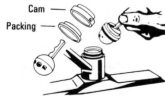

Cam
Packing

2. Slip packing and then cam on new ball and insert into socket. Reassemble faucet in reverse order.

NOTE: After any service is completed, remove aerator and flush hot and cold lines.

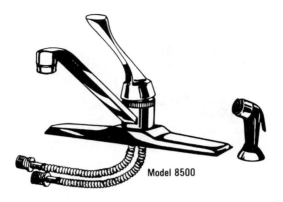

Model 8500

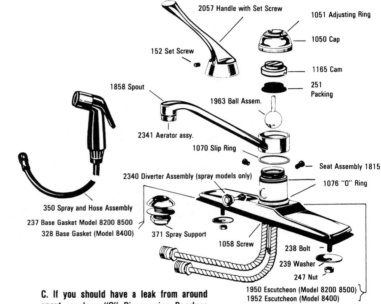

2057 Handle with Set Screw
152 Set Screw
1858 Spout
2341 Aerator assy.
1070 Slip Ring
2340 Diverter Assembly (spray models only)
350 Spray and Hose Assembly
237 Base Gasket Model 8200 8500
328 Base Gasket (Model 8400)
371 Spray Support
1058 Screw
1051 Adjusting Ring
1050 Cap
1165 Cam
251 Packing
1963 Ball Assem.
Seat Assembly 1815
1076 "O" Ring
238 Bolt
239 Washer
247 Nut
1950 Escutcheon (Model 8200 8500)
1952 Escutcheon (Model 8400)

C. If you should have a leak from around spout—replace "O" Rings using Peerless repair kit #2078.

1. Shut off water. Remove handle, cap assembly and lift out ball assembly.

2. Gently rotate spout and lift off.

"O" rings

3. Cut and remove old "O" Rings. Stretch new "O" Rings and snap into grooves on body. Reassemble faucet in reverse order. Be careful when putting spout back on to gently rotate until it rests on plastic ring.

A. If you should have a leak from under handle—

1. Do not shut off water. Remove handle and use Peerless wrench to tighten adjusting ring.

2. Tighten adjusting ring until no water will leak around stem when faucet is on and pressure is exerted to force ball into socket.

D. If you should have problem with spray attachment—

1. Shut off water. Remove handle, cap assembly and ball assembly.

2. Gently rotate spout and lift off.

3. Remove diverter assembly by pulling out with finger. Wash diverter thoroughly and replace. If diverter assembly appears damaged, write Peerless Faucet for RP No. 2340 and cost.

Repair instructions for Peerless single-handle kitchen faucet.

Courtesy Peerless Faucets

PEERLESS TWO HANDLE (DURALAC) KITCHEN FAUCETS

To Disassemble—shut off water.

1. Pry off handle button, remove screw and lift off handle.

2. Unscrew bonnet counter-clockwise.

A. If you should have a leak from spout outlet—install new seats and springs using Peerless Repair Kit #1815.

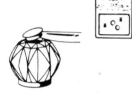

1. Remove handle and bonnet.

2. Lift out stem unit.

3. Remove old seats and springs and insert new ones. Reassemble faucet in reverse order.

If leak persists—Use Peerless repair kit #2083 to install new stem unit.

1. Remove handle and bonnet.

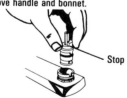

        —Stop

2. Lift out stem and replace with new stem unit.

Note: For **knob** handle faucets, both stops must point toward spout. For **blade** handle faucets, left hand stop points toward spout, right hand stop, points away from spout.

3. Reassemble faucet in reverse order.

If you should have a leak from top or bottom of spout body, replace "O" Rings, using Peerless repair kit #2080.

1. Remove spout nut, gently rotate spout and lift off.

2. Cut and remove old "O" Rings. Stretch new "O" Rings and snap into grooves on body. Reassemble faucet in reverse order. Be careful when putting spout back on to gently rotate until it rests on plastic ring.

If you should have a problem with spray attachment—

1. Remove spout nut.

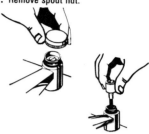

2. Back diverter out with screwdriver.

3. Replace spout nut and flush faucet.

4. Wash off and replace diverter. Replace spout nut, tighten lightly.

Be sure to flush both hot and cold water lines for one minute each before replacing aerator (diverter, if spray model).

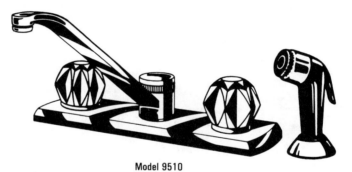

Model 9510

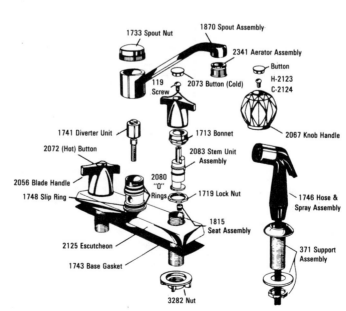

1733 Spout Nut
1870 Spout Assembly
2341 Aerator Assembly
Button
H-2123
C-2124
119 Screw
2073 Button (Cold)
2067 Knob Handle
1741 Diverter Unit
1713 Bonnet
2072 (Hot) Button
2083 Stem Unit Assembly
2056 Blade Handle
2080 "O" Rings
1719 Lock Nut
1746 Hose & Spray Assembly
1748 Slip Ring
1815 Seat Assembly
2125 Escutcheon
371 Support Assembly
1743 Base Gasket
3282 Nut

Courtesy Peerless Faucets

Repair instructions for Peerless two-handle kitchen faucet.

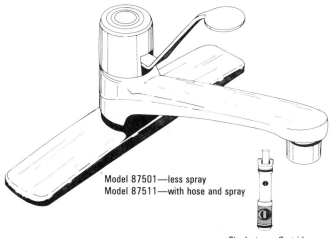

Model 87501—less spray
Model 87511—with hose and spray

Single-Lever Cartridge

Courtesy Stanadyne/Moen Division

Fig. 2-27. Moen single-lever kitchen faucet.

Maintenance: Your new Touch Control by MOEN faucet will give good service for years with a minimum of maintenance. If your faucet drips from the spout or leaks up through the handle, replace the cartridge, following the instructions given below. If it leaks from the top or bottom of the spout hub, replace the spout O-rings. If your spray model will not completely divert water from the spout to the spray head, replace the diverter. If water flow is weak or irregular, unscrew the aerator, clean and replace.

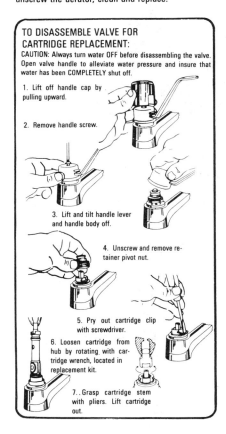

TO DISASSEMBLE VALVE FOR CARTRIDGE REPLACEMENT:

CAUTION: Always turn water OFF before disassembling the valve. Open valve handle to alleviate water pressure and insure that water has been COMPLETELY shut off.

1. Lift off handle cap by pulling upward.

2. Remove handle screw.

3. Lift and tilt handle lever and handle body off.

4. Unscrew and remove retainer pivot nut.

5. Pry out cartridge clip with screwdriver.

6. Loosen cartridge from hub by rotating with cartridge wrench, located in replacement kit.

7. Grasp cartridge stem with pliers. Lift cartridge out.

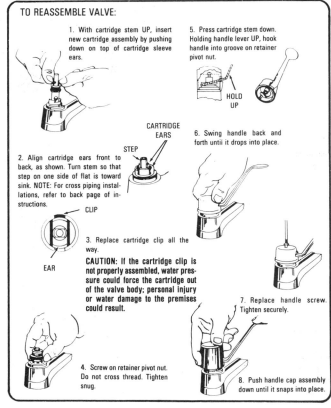

TO REASSEMBLE VALVE:

1. With cartridge stem UP, insert new cartridge assembly by pushing down on top of cartridge sleeve ears.

2. Align cartridge ears front to back, as shown. Turn stem so that step on one side of flat is toward sink. NOTE: For cross piping installations, refer to back page of instructions.

3. Replace cartridge clip all the way.

CAUTION: If the cartridge clip is not properly assembled, water pressure could force the cartridge out of the valve body; personal injury or water damage to the premises could result.

4. Screw on retainer pivot nut. Do not cross thread. Tighten snug.

5. Press cartridge stem down. Holding handle lever UP, hook handle into groove on retainer pivot nut.

HOLD UP

CARTRIDGE EARS

STEP

CLIP

EAR

6. Swing handle back and forth until it drops into place.

7. Replace handle screw. Tighten securely.

8. Push handle cap assembly down until it snaps into place.

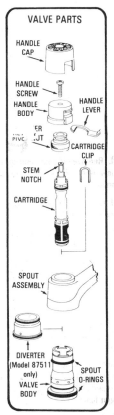

VALVE PARTS

HANDLE CAP

HANDLE SCREW

HANDLE BODY

HANDLE LEVER

PIVOT

CARTRIDGE CLIP

STEM NOTCH

CARTRIDGE

SPOUT ASSEMBLY

DIVERTER (Model 87511 only)

VALVE BODY

SPOUT O-RINGS

Stanadyne/Moen Division

Repair instructions for Moen single-lever kitchen faucet.

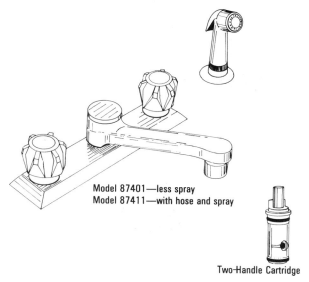

Model 87401—less spray
Model 87411—with hose and spray

Two-Handle Cartridge

Courtesy Stanadyne/Moen Division

Fig. 2-28. Moen two-handle kitchen faucet.

Maintenance: Your new Touch Control by MOEN faucet will give good service for years with a minimum of maintenance. If your faucet drips from the spout or leaks up through the handles, replace the cartridge, following the instructions given below. If it leaks from the top or bottom of the spout hub, replace the spout O-rings. If water flow is weak or irregular, unscrew the aerator, clean and replace.

TO DISASSEMBLE VALVE FOR CARTRIDGE REPLACEMENT:

CAUTION: Always turn water OFF before disassembling the valve. Open valve handles to alleviate water pressure and insure that water has been COMPLETELY shut off.

1. Pry off handle cover with a thin-bladed instrument.

2. Remove handle screw.

3. Pull handle knob up and off.

4. Unscrew cartridge nut with pliers and remove it.

5. Grasp cartridge stem with pliers. Pulling straight up, lift cartridge up.

TO REASSEMBLE VALVE:

IMPORTANT: Follow these instructions carefully to prevent irreparable damage to cartridge.

1. Turn cartridge stem counterclockwise, to "ON" position, so that waterway holes in body of cartridge are lined up.

4. Tighten firmly with pliers.

2. Push the cartridge straight down into the body. The key on the cartridge should be fitted into the notch on the body. This will position the handle stop on the cartridge to the side, as shown.

CAUTION: Make sure the other cartridge on the opposite side is completely assembled into the faucet.

5. Drop handle down on the stem, making sure that flats in the handle knob are matched with the flats on the cartridge stem. Turn water supplies on. Test operation of faucet.

3. Screw down the cartridge nut by hand. DO NOT CROSS THREADS.

6. Replace handle screw with screwdriver. Tighten securely.

7. With knob in OFF position, press handle cover in place.

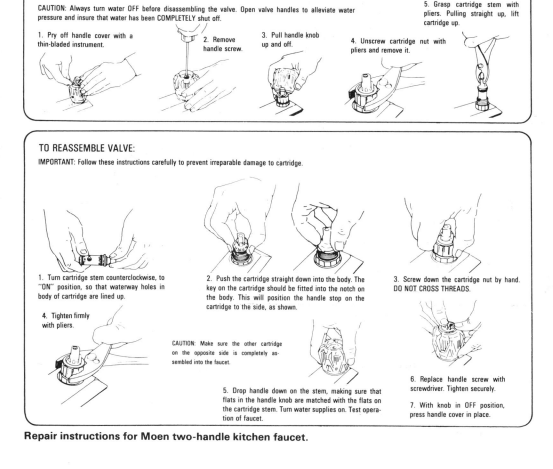

Repair instructions for Moen two-handle kitchen faucet.

VALVE PARTS

HANDLE COVER

HANDLE SCREW

HANDLE KNOB

CARTRIDGE NUT

CARTRIDGE

BODY

FAUCET PARTS

SPOUT CAP

DIVERTER (Model 87411)

SPOUT SEALS

Stanadyne/Moen Division

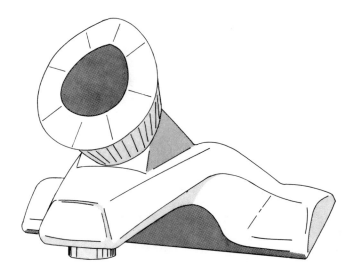

Courtesy Stanadyne/Moen Division

Fig. 2-29. Moen single-handle lavatory faucet.

Maintenance: Your new Touch Control by MOEN faucet will give good service for years with a minimum of maintenance. If your faucet drips from the spout or leaks up through the handles, replace the cartridge, following the instructions given below. If water flow is weak or irregular, unscrew the aerator, clean and replace.

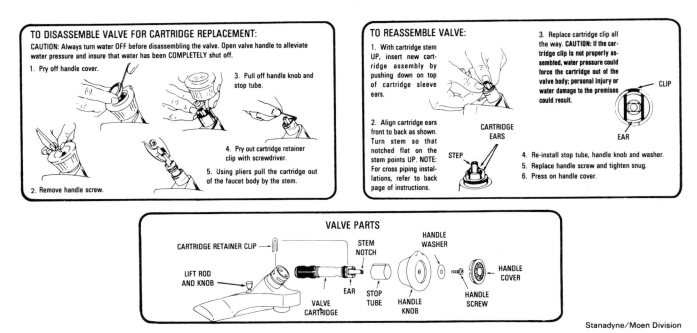

Stanadyne/Moen Division

Repair instructions for Moen single-handle lavatory faucet.

Courtesy Stanadyne/Moen Division

Fig. 2-30. Moen two-handle lavatory faucet.

Maintenance: Your new Touch Control by MOEN faucet will give good service for years with a minimum of maintenance. If your faucet drips from the spout or leaks up through the handle, replace the cartridge, following the instructions given below. If water flow is weak or irregular, unscrew the aerator, clean and replace.

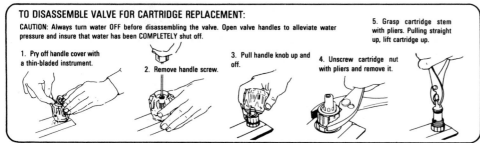

TO DISASSEMBLE VALVE FOR CARTRIDGE REPLACEMENT:

CAUTION: Always turn water OFF before disassembling the valve. Open valve handles to alleviate water pressure and insure that water has been COMPLETELY shut off.

5. Grasp cartridge stem with pliers. Pulling straight up, lift cartridge up.

1. Pry off handle cover with a thin-bladed instrument.

2. Remove handle screw.

3. Pull handle knob up and off.

4. Unscrew cartridge nut with pliers and remove it.

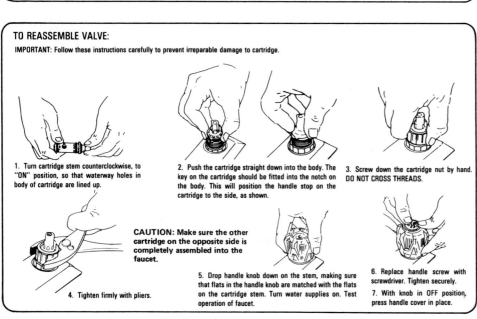

TO REASSEMBLE VALVE:

IMPORTANT: Follow these instructions carefully to prevent irreparable damage to cartridge.

1. Turn cartridge stem counterclockwise, to "ON" position, so that waterway holes in body of cartridge are lined up.

2. Push the cartridge straight down into the body. The key on the cartridge should be fitted into the notch on the body. This will position the handle stop on the cartridge to the side, as shown.

3. Screw down the cartridge nut by hand. DO NOT CROSS THREADS.

CAUTION: Make sure the other cartridge on the opposite side is completely assembled into the faucet.

4. Tighten firmly with pliers.

5. Drop handle knob down on the stem, making sure that flats in the handle knob are matched with the flats on the cartridge stem. Turn water supplies on. Test operation of faucet.

6. Replace handle screw with screwdriver. Tighten securely.

7. With knob in OFF position, press handle cover in place.

Repair instructions for Moen two-handle lavatory faucet.

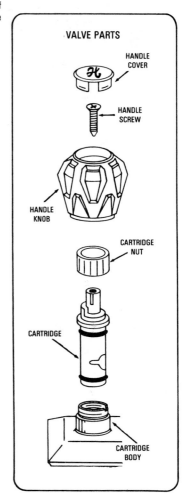

VALVE PARTS

HANDLE COVER

HANDLE SCREW

HANDLE KNOB

CARTRIDGE NUT

CARTRIDGE

CARTRIDGE BODY

Courtesy Stanadyne/Moen Division

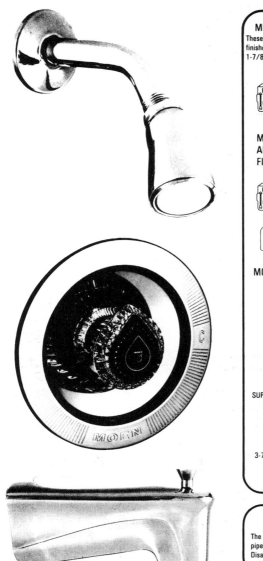

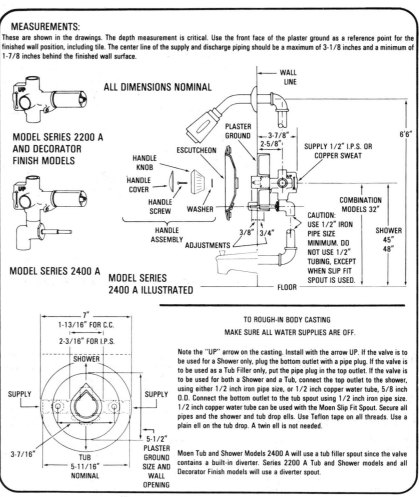

MEASUREMENTS:

These are shown in the drawings. The depth measurement is critical. Use the front face of the plaster ground as a reference point for the finished wall position, including tile. The center line of the supply and discharge piping should be a maximum of 3-1/8 inches and a minimum of 1-7/8 inches behind the finished wall surface.

ALL DIMENSIONS NOMINAL

WALL LINE

MODEL SERIES 2200 A AND DECORATOR FINISH MODELS

PLASTER GROUND
3-7/8"
2-5/8"
6'6"
ESCUTCHEON
HANDLE KNOB
SUPPLY 1/2" I.P.S. OR COPPER SWEAT
HANDLE COVER
COMBINATION MODELS 32"
HANDLE SCREW
WASHER
CAUTION: USE 1/2" IRON PIPE SIZE MINIMUM. DO NOT USE 1/2" TUBING, EXCEPT WHEN SLIP FIT SPOUT IS USED.
SHOWER 45" 48"
HANDLE ASSEMBLY
3/8" 3/4"
ADJUSTMENTS
FLOOR

MODEL SERIES 2400 A

MODEL SERIES 2400 A ILLUSTRATED

7"
1-13/16" FOR C.C.
2-3/16" FOR I.P.S.
SHOWER
SUPPLY
SUPPLY
5-1/2"
PLASTER GROUND SIZE AND WALL OPENING
3-7/16"
TUB
5-11/16" NOMINAL

TO ROUGH-IN BODY CASTING

MAKE SURE ALL WATER SUPPLIES ARE OFF.

Note the "UP" arrow on the casting. Install with the arrow UP. If the valve is to be used for a Shower only, plug the bottom outlet with a pipe plug. If the valve is to be used as a Tub Filler only, put the pipe plug in the top outlet. If the valve is to be used for both a Shower and a Tub, connect the top outlet to the shower, using either 1/2 inch iron pipe size, or 1/2 inch copper water tube, 5/8 inch O.D. Connect the bottom outlet to the tub spout using 1/2 inch iron pipe size. 1/2 inch copper water tube can be used with the Moen Slip Fit Spout. Secure all pipes and the shower and tub drop ells. Use Teflon tape on all threads. Use a plain ell on the tub drop. A twin ell is not needed.

Moen Tub and Shower Models 2400 A will use a tub filler spout since the valve contains a built-in diverter. Series 2200 A Tub and Shower models and all Decorator Finish models will use a diverter spout.

FLUSHING NOTE

The valve body and supplies should be flushed before placing them into service to remove pipe chips, scale or other foreign material left in pipes after installation. These might clog or damage the valve and cause a leak. To do this, make sure all water supplies are OFF. Disassemble as shown on next page. Slowly turn on both hot and cold water supplies and thoroughly flush out the body and lines.

Courtesy Stanadyne/Moen Division

Fig. 2-31. Moen single-handle tub/shower valve.

Maintenance: Your new Touch Control by MOEN faucet will give good service for years with a minimum of maintenance. If your faucet drips from the spout or showerhead, replace the cartridge, following the instructions given below.

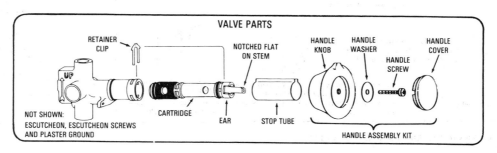

VALVE PARTS

RETAINER CLIP
NOTCHED FLAT ON STEM
HANDLE KNOB
HANDLE WASHER
HANDLE COVER
HANDLE SCREW
NOT SHOWN: ESCUTCHEON, ESCUTCHEON SCREWS AND PLASTER GROUND
CARTRIDGE
EAR
STOP TUBE
HANDLE ASSEMBLY KIT

TO DISASSEMBLE VALVE FOR CARTRIDGE REPLACEMENT:

Caution: Always turn water OFF before disassembling the valve. Open valve handle to alleviate water pressure and insure that water has been COMPLETELY shut off.

1. Pry off handle cover.

2. Remove handle screw.

3. Pull off handle knob and stop tube.

4. Pry out cartridge retainer clip with screwdriver.

5. Loosen cartridge from hub by rotating with cartridge wrench, located in replacement kit.

6. Using pliers, pull the cartridge out of the faucet body by the stem.

TO REASSEMBLE VALVE:

1. With cartridge stem OUT, insert new cartridge assembly by pushing in on cartridge sleeve ears.

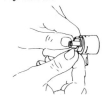

2. Align cartridge ears top to bottom as shown. Turn stem so that notched flat on the stem points UP. NOTE: For cross piping installations, refer to back page of instructions.

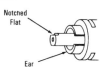

Notched Flat
Ear

3. Replace cartridge clip all the way. CAUTION: If the cartridge clip is not properly assembled water pressure could force the cartridge out of the valve body; personal injury or water damage to the premises could result.

Clip

Ear

4. Reinstall stop tube, handle knob and washer.

5. Replace handle screw and tighten snug.

6. Press on handle cover.

Stanadyne/Moen Division

Repair instructions for Moen single-handle tub/shower faucet.

CHAPTER 3

Repairing Toilet Tanks

Men have often had a knack for making simple things complicated. Most of the toilet tanks in use today are as outdated as the Model-T Ford. Everything considered, the maze of wires, floats, rods, levers, tubes, guides, and balls that comprises the working parts of the average water closet tank work amazingly well. Repair or replacement of some or all of the parts of a water closet tank becomes necessary when corrosion, mineral build-up, or normal wear cause these parts to malfunction. The replacement of old-fashioned tank parts with the newer parts shown in this chapter is strongly recommended.

In order to make it easier to repair a water closet tank, let us study the chain of events set in action when the trip lever of the tank is pushed down. Refer to Fig. 3-1 as we go along. Trip lever (A) pulls up on the upper pull wire (B), which is linked to the lower pull wire (C). The lower pull wire, sliding through the guide (D), is screwed into the top of the tank ball (E). The tank ball, upon being pulled up, becomes free of the water pressure that had been holding it in place on the seat of the flush valve (F) and becomes water-borne. The water then starts to rush out of the tank and into the closet bowl. As the water level in the tank drops, the float ball (G) connected to the ballcock (H) by float rod (I) through compound lever (J) also drops with the water level. As the water level drops, the weight of the float ball and float rod exerts pressure to raise the plunger (K) off the ballcock washer seat (L).

The ballcock (which is a water valve) is thus turned on and water begins to flow into the tank. The water flowing into the tank is not coming in as fast as the water already there is leaving; therefore, the tank ball drops with the receding water and is guided into proper contact with the flush valve seat by the guide. When it is properly seated, the tank ball prevents any more water from leaving the tank. The float ball is now riding the crest of the incoming water, and when the water level has reached the water line (O), the float exerts pressure on the plunger through the float rod and the compound lever, forcing the plunger into contact with the ballcock washer seat. This action of closing the valve in the ballcock shuts off the water entering the tank. Now that we know the sequence of events that flushes out the water closet bowl, let us see some of the things that commonly go wrong in the operation of the water closet tank.

PROBLEM: The ballcock will not shut off, the water level rises to high, and water runs out of the tank through the overflow tube.

CURE: (1) The float ball may be rubbing on the side of the closet tank and the ball cannot rise with the incoming water; in this case, bend the float rod horizontally until the float rides free on the water.

(2) The float ball may have developed a leak and lost its bouyancy. In this case, close (shut off) the water valve on the pipe connecting to the ballcock. This valve should be located below the left side of the

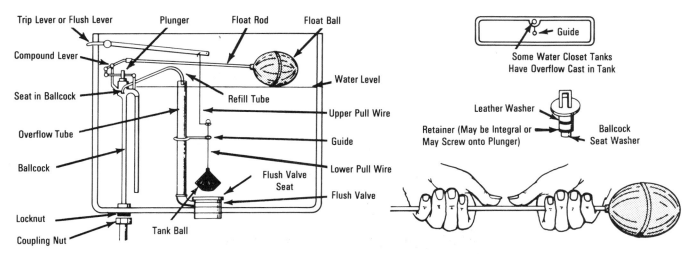

Fig. 3-1. Old-fashioned tank components.

water closet tank. Unscrew the float ball from the float rod by turning the ball counterclockwise; when the float ball is free, shake it; if there is water in the ball, it must be replaced.

(3) Replace the ballcock washer. The ballcock washer is on the bottom of the plunger.

PROBLEM: You have a very high water bill; you suspect a leak in the water closet tank, but you cannot locate it.

CURE: Pour a tablespoon of dark blue or black ink or food coloring into the water closet tank. Wait for 3 or 4 minutes, then observe the water in the closet bowl. If the water in the bowl has turned dark, there is a leak in the closet tank where the tank ball seats on the flush valve, or there is a leak in the overflow tube. Replace the tank ball first, preferably with one of the new "flapper-type" tank valves, furnished as a complete kit that includes the flapper-type valve, a stainless steel seat, and the necessary cement, as shown in Fig. 3-10.

If replacing the tank ball does not stop the leak, the leak is in the overflow tube. If the overflow tube is made of brass, use slip-joint pliers to grasp it just above the connection to the flush valve. Turn the overflow tube counterclockwise to unscrew it from the flush valve. If the overflow tube is made of plastic, the complete flush valve should be replaced.

PROBLEM: The closet tank does not refill after the closet bowl has been flushed.

CURE: (1) The float rod may be bent, causing the float ball to drag on the side of the tank. If this happens, the float ball cannot drop and the water will not be turned on by the ballcock. Bend the rod horizontally to free the float ball from the side of the tank.

(2) The tank ball does not drop into position to seat properly on the flush valve. The lower pull wire

slides through the eye of the guide and is screwed into the tank ball. Position the guide so that the eye is centered on the flush valve opening. Refer to the closet tank illustration for measurements. Be certain that the upper and lower pull wires are installed as they are pictured in the illustration and that these wires are not bent. The upper and lower pull wires must be straight and guided correctly for the tank ball to drop into and seat properly on the flush valve.

PROBLEM: The trip lever must be depressed several times in order to flush the closet bowl.

CURE: This is usually due to too much play in the trip lever itself. Install a new trip lever. Also, check for proper installation of the upper and lower pull wires.

HOW TO ADJUST THE WATER LEVEL IN THE TANK

Some ballcocks have an adjusting screw that raises and lowers the float rod and thus adjusts the water level in the tank. This screw will be located at the float rod connection to the ballcock. Another way to raise or lower the water level in the tank is to bend the float rod (I) in Fig. 3-1. Hold the float rod firmly with the left hand. Using the right hand, bend the rod firmly with the left hand. Using the right hand, bend the rod up to raise the water level in the tank; bend the rod down to lower the water level. The correct water level is approximately one inch below the top of the overflow tube. Fig. 3-1 shows the old standard types of ballcock, float ball and float rod, flush valve, upper and lower pull wires, and tank ball installed in the water closet tank. As mentioned earlier, the use of newer type of fill valves and "flapper-type" tank balls is strongly recommended.

WATER CLOSET FILL VALVES

Two types of water closet fill valves that are superior to the old-fashioned ballcock in design and performance are the *Fillmaster* and the *Fluidmaster*.

The Fillmaster Fill Valve

The *Fillmaster*, shown in Figs. 3-2 and 3-3, uses the hydraulic force of the supply water to provide positive opening and closing of the fill valve. Using no float rod or float ball, the *Fillmaster* measures the water level in the tank by means of a diaphragm that responds to the head (or height) of water in the closet tank. The water level in the tank can be adjusted to the desired level by turning the knob marked "ADJ." Each full turn of the knob changes the water level by one inch. The *Fillmaster* fill valve modulates the flow of water into the tank. Water seepage through the flush valve is replaced with no cycling noise. The valve is constructed of plastic, rubber, and stainless steel. An antisiphon version of the *Fillmaster* is available for use in cities where plumbing codes require back-siphonage prevention. A

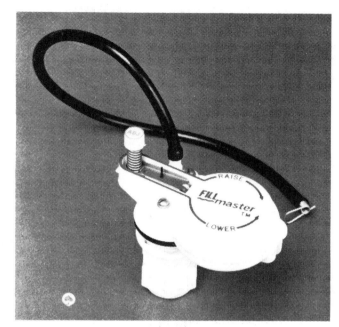

Courtesy J.H. Industries, Inc.

Fig. 3-2. The *Fillmaster* valve unit.

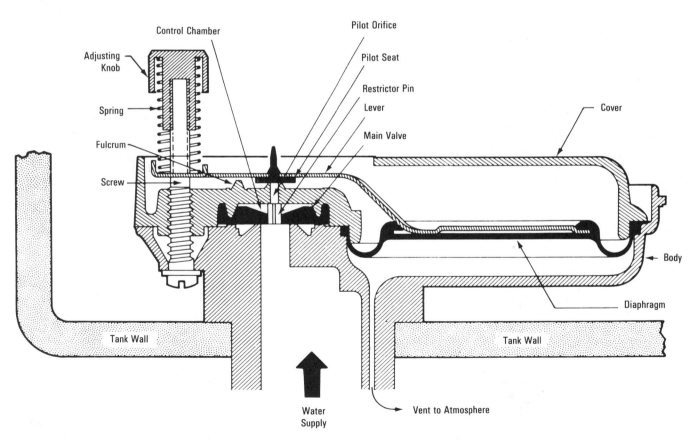

Courtesy J.H. Industries, Inc.

Fig. 3-3. Cutaway view of the *Fillmaster* valve.

"flapper-type" tank ball is included in a *Fillmaster* kit to completely modernize the major control devices of a water closet tank all at one time. See Figs. 3-4 through 3-8 for installation and operation sequence of the *Fillmaster*.

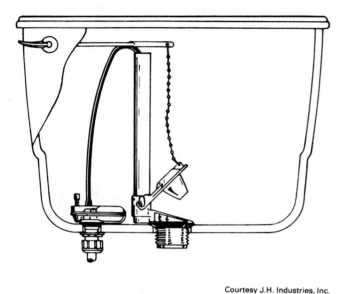

Courtesy J.H. Industries, Inc.

Fig. 3-4. Typical *Fillmaster* installation.

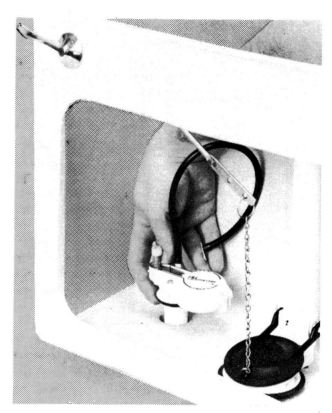

Courtesy J.H. Industries, Inc.

Fig. 3-6. Inserting *Fillmaster* valve in tank.

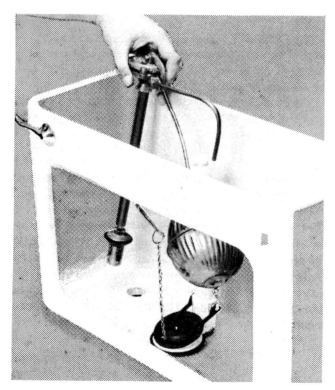

Courtesy J.H. Industries, Inc.

Fig. 3-5. Removing old-fashioned tank ballcock.

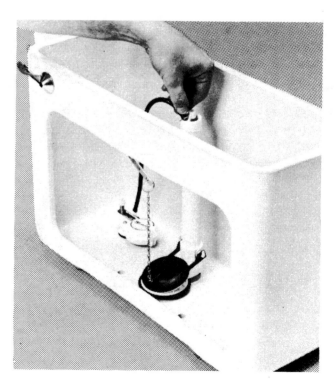

Courtesy J.H. Industries, Inc.

Fig. 3-7. Clipping rubber refill tube to overflow pipe.

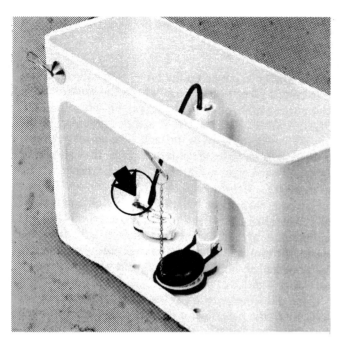

Fig. 3-8. Water-level adjustment knob.

The Fluidmaster Fill Valve

The *Fluidmaster* fill valve uses no float ball or float rod to actuate the valve. The valve redirects the water pressure to cause the shut-off action to take place when the water level in the tank reaches the desired height. A stainless steel seating surface is used at the shut-off point. The water level in the tank is regulated by a sliding clip.

The *Fluidmaster* operates in a full-on or full-off position, there is no modulation in its action. Water leakage at the tank ball, through the flush valve, is signalled by the sound of the valve refilling the tank, when the tank has not been flushed.

In addition to the Model 100 and Model 200, the *Fluidmaster* is also available in an antisiphon Model 400 for use in cities where the plumbing code requires back-siphonage protection (see Fig. 3-9). The *Fluidmaster Flusher-Fixer* kit, a "flapper-type" tank ball with a stainless steel seating surface, can be installed to modernize the working parts of the water closet tank (see Fig. 3-10).

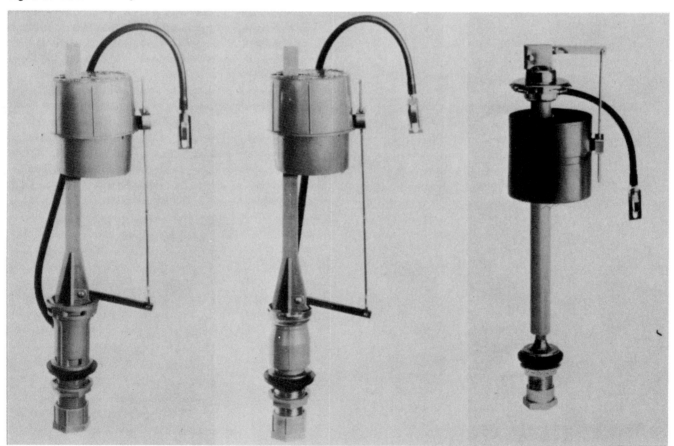

Fig. 3-9. *Fluidmaster* **fill valves.**

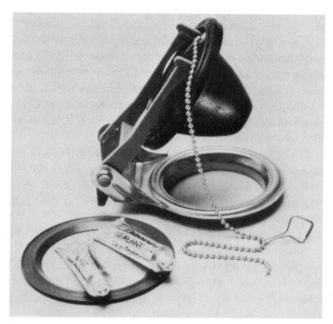

Courtesy Fluidmaster, Inc.

Fig. 3-10. *Fluidmaster Flusher-Fixer* kit.

CHAPTER 4

Repairing a
Trip-Lever Bath Drain

The trip-lever bath drain shown in Fig. 4-1 is subject to two common problems. When the raised portion on the back of the faceplate (over which the lever must ride) becomes worn, the lever will no longer stay in a set position. Also, if the lever is not used and is allowed to stay in one position, soap, grime, hair, etc., present in the bath water will build up around the plug, making it immovable. The worn trip-lever faceplate can be replaced; new ones are available in plumbing supply departments, but in order to replace it, the linkage and plug attached to the faceplate must be removed through the overflow piping. Loosen and remove the two screws securing the faceplate. Pull the faceplate out and up; the linkage and plug should come out with the faceplate. If the linkage and plug cannot be pulled up, spray a good quality penetrating oil, *Liquid Wrench* or *WD 40*, into the overflow piping. The oil will run down the piping and around the plug and will usually free it in a short time. You may have to repeat the spraying two or three times over a period of several hours to allow the oil to do its work. When the plug and linkage have been removed, use long-nose pliers to remove the cotter pin securing the faceplate to the clevis. Discard the old faceplate and connect the new one to the linkage using the old cotter pin. Clean the outside surfaces of the plug, using fine sandpaper or steel wool, and grease the plug lightly before inserting it and the linkage into the overflow piping. Secure

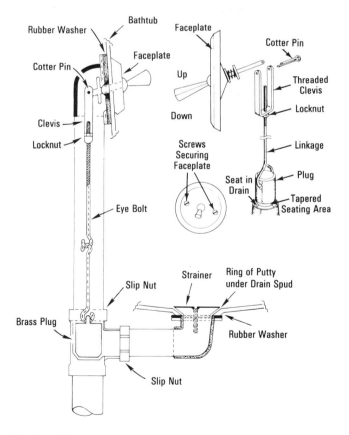

Fig. 4-1. A brass-plug-type connected waste and overflow fitting.

71

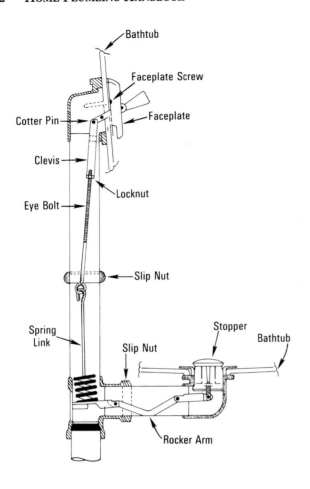

Fig. 4-2. A rocker-arm-type connected waste and overflow fitting.

the faceplate to the overflow piping and test the trip lever for proper operation.

As shown in Fig. 4-1, when the lever is in DOWN position, the plug is lifted off of the drain seat. When the lever is in UP position, the plug should be positioned on the drain seat, thus holding the water in the tub. If the linkage was correctly adjusted before changing the faceplate, no additional adjustment should be necessary. Push the trip lever to the UP position, if it will not go completely up, the linkage rod must be shortened. Remove the faceplate and lift out the rod and plug. Loosen the locknut and turn the linkage rod into the clevis three or four turns, then tighten the locknut and repeat the test procedure. When the trip lever is in the UP position and will not hold the water in the tub, the rod must be lengthened by turning it out of the clevis. It may be necessary to repeat this process several times until the correct adjustment is made and the trip lever works properly.

The rocker-arm-type trip-lever drain shown in Fig. 4-2 is also adjusted by turning the linkage rod into or out of the clevis and securing it with the locknut. The linkage and spring can be removed by removing the faceplate screws and pulling the faceplate up and out. The stopper can be adjusted up or down by turning it clockwise or counterclockwise. The rocker arm may become corroded and often fails to operate properly due to hair, soap, etc., which collect at the spring link. The remedy for this is to remove the linkage and spring and clean them. The rocker arm and stopper can be pulled up and out of the tub opening when cleaning or adjustment is necessary.

CHAPTER 5

How to Deal with Stopped-up Drains

Grease, hair, and soap build-up over a period of time in drain piping are the most common causes of drain stoppages inside the home. Roots and broken drain tile are responsible for most stoppages in the sewer from the home to the main sewer in the street or alley. If all the drains in a home run slow or are completely stopped up, it is safe to assume that the sewer from the home to the main sewer is blocked. Unblocking it is not a job for the average householder. The equipment necessary is cumbersome, heavy, and requires experience in its use. The operator has to feel his way as he works the cutter through the sewer. He knows from the way the equipment handles when he has encountered tree roots or a broken tile. If all the drains in your home are stopped up, leave this job to the experts and call your local plumber or a company specializing in drain cleaning. After the roots have been removed, a copper sulphate solution should be poured into the sewer periodically, preferably through a clean-out opening, to slow down or prevent regrowth of roots. Copper sulphate solutions are sold for this purpose by plumbing shops and hardware stores.

Industrial types of chemical drain cleaners used by plumbers are very effective, but they usually contain very strong acids and should be used *only* by skilled professionals.

KITCHEN SINK DRAINS

A plunger is often very effective in unstopping sink drains. When a plunger is pushed down over the drain, a shock wave is transmitted through the water in the drain, and steady use of the plunger, causing a series of shock waves, will often break a stoppage loose. If you have a two-compartment sink, use one plunger to seal the opening on one side of the sink while using another plunger on the other side. If the plunger does not open the drain, it will be necessary to run a cable through the drain piping to get to the stoppage. A cleanout plug may be installed in the drain piping under the sink for this purpose. If there is no cleanout plug, disconnect the sink trap and insert the cable into the drain piping at this point. Refer to Fig. 5-1.

If there is water in the sink, above the cleanout plug or above the trap, dip or sponge the water out or place a large pan or bucket under the trap before removing the trap to catch any water in the sink or in the piping. Insert the cable into the drain piping; rotating the cable (spinning the "top") will help to work the cable around turns or fittings in the piping. Working the cable forward and back will help to break through the obstruction. After reaching and breaking through the obstruction, remove the cable, screw the cleanout plug

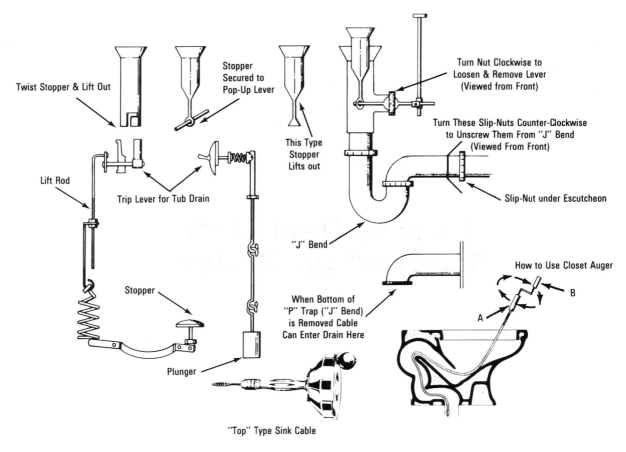

Fig. 5-1. Using spring cables to open drains.

into the cleanout or reconnect the trap, and run water into the drain to make certain the drain is open. When it is open, turn the hot-water faucet on for three or four minutes. The hot water will melt the congealed grease in the drain piping.

LAVATORY DRAINS

The procedure for unstopping a lavatory drain is basically the same as unstopping a kitchen sink drain. The most common cause of a stopped-up lavatory drain is an accumulation of hair. If the lavatory has a pop-up drain, lift out the stopper in the drain. Some stoppers have a lock-in device, i.e., they are secured to the lever which is connected to the lift rod. Other stoppers twist and lift out. Hair will often catch on the bottom of the stopper and cause the lavatory drain to stop up. If this is the problem, removing the stopper and cleaning out the accumulation of hair will open the drain.

The plunger is a tool that also is very useful for unstopping lavatory drains. The lavatory has a drain connection and also an overflow connection. The overflow opening may be on either the front or the back side of the bowl, near the top. Block the overflow connection with a wet cloth (or a small plunger) when plunging the drain connection. If the drain piping is stopped up, use a smooth-jaw monkey wrench or *Ridgid* offset hex wrench to turn the slip-nuts on the trap counterclockwise to loosen. When the slip-nuts are loose, slide the "J" bend or bottom of the "P" trap down off of the drain fitting. Slide the escutcheon forward, away from the wall, to expose the trap connection at the wall. If the trap has a slip-nut connection at the wall, unscrew the slip-nut and pull the balance of the trap out of the drain piping.

If the trap is soldered into a lead pipe at the wall or has a threaded connection (other than a slip-nut), it would be best to leave the rest of the trap connected at the wall and thread a small sink cable (1/4" or 3/8") into the drain piping through the trap connection, as illustrated. Spin the cable as it works its way through the drain piping; this will help to work the cable around fittings in the piping, and the spinning action will also help clean the drain line. The top-type sink cable shown in Fig. 5-1 is very useful for unstopping lavatory drains.

BATHTUB DRAINS

The cause of most stopped-up bath drains is also hair. Quite often hair will collect around the pop-up drain parts, a spring, a plunger, or a stopper. If an accumulation of hair has caused the stopped-up drain, taking the pop-up parts out and cleaning them will usually open the drain. To disassemble a typical pop-up drain, remove the screws securing the overflow plate. Pull the overflow plate forward and upward, and work the lift rod out. Remove any hair, soap, or grease from the spring or plunger. The stopper in the drain outlet of the tub can be removed by pulling up on the stopper and working it out of the drain opening. Remove any accumulation of hair, soap, or grease from the stopper. If after cleaning the plunger, spring, or stopper, the bath drain is still stopped, stuff a wet rag into the overflow pipe to block it and use a plunger on the drain opening. This plunging action will usually open the drain.

If using the plunger on the drain fails to open the drain, then a small cable, such as the top-type cable, should be used. Feed the cable into the drain at the overflow connection, rotating the drum will help the cable to work its way around the fittings in the bath trap. When the drain is open, reassemble the drain connections.

TOILET BOWLS

The most common cause for a stopped-up toilet bowl is a sanitary napkin carelessly disposed of in the toilet bowl. The next common troublemakers are combs, toys, or pencils. A large plunger, especially made for plunging toilet bowls, will usually force sani-tary napkins and accumulated wastes on through the bowl. Toys, especially soft plastic toys that bend, can usually be forced through a closet bowl with a good plunger. Straight objects, pencils, and combs will often be washed half-way through the trap but will catch or jam in the final turn of the trap. The only practical way to dislodge a pencil, comb, or similar straight object is to use a closet auger. If a plunger fails to unstop a closet bowl, see Fig. 5-1 and use a closet auger in the following manner:

Grasp the closet auger at point (A) with one hand and the handle (B) with the other hand. Pull the handle up and out, until the spring end stops against the end of the guide tube. Insert the spring end of the auger into the entrance to the trap in the closet bowl. Push down on the handle of the auger, forcing the spring into the closet bowl, and turn the handle in a clockwise direction while forcing the spring into the bowl. When the handle has come to rest against the guide tube, turn the handle clockwise three or four times. This action will help to break up or catch the offending article in the spring end of the auger. Pull the handle up and out, and repeat the process again if necessary. A word of caution: Toilet bowls are made of china and are easily breakable; turning the auger handle while inserting the spring end into the toilet bowl will minimize the chances of breaking the toilet bowl.

A heavy-duty plunger is shown in Fig. 1-18.

The tools and techniques listed in this chapter will enable you to handle most stopped-up drains. There are times, however, when special heavy-duty equipment is needed to unstop drains. If you've made every effort to do it yourself without success, call one of the companies specializing in this work. They have the equipment, with special types of cutters, etc., to handle difficult jobs.

CHAPTER 6

How to Work with Copper Tubing

Joining two or more pieces of copper tubing or repairing a leak in copper tubing is not a difficult operation. There is a very simple way to make repairs or add pipe and fittings to an existing piping system, which I will describe later in this chapter.

The most common and least expensive method to make repairs to, or changes in, a copper piping system is to solder (or "sweat") the joints. There are a few simple rules for soldering copper joints. If you will learn and follow these rules, you can make perfect solder joints every time.

1. The male end of the piping and the female end of the fitting to be joined together must be clean and bright.
2. Heat must be applied at the right place on the fitting. Capillary action will then pull the solder into the joint.
3. Use plain 50/50 wire solder and a noncorrosive solder paste. *DO NOT USE* acid-core solder or rosin-core solder.
4. The piping being joined *must* be dry. It is impossible to solder a joint properly with water in the pipe. You may find that a valve will not shut off completely and allows a trickle of water to flow through the pipe. You can stop this water flow long enough to solder a joint in the pipe in the following manner: Obtain a slice of bread and break it into small pieces. Insert these pieces of bread into the pipe and pack it tightly, using the

eraser end of a pencil as a tamping tool; this will stop the flow of water and enable you to solder the joint. When the solder joint has been made and the water pressure is turned on in the pipe, the "bread plug" will literally disintegrate and can be washed out of the pipe through an opened valve or faucet. If your faucets have aerators on the spouts, remove the aerators while flushing out the piping.

5. If you are working on a closed system of piping, open a valve or turn on a faucet as shown in Fig. 6-1.

A closed system of water piping is connected to a water supply line at one end and to valves, fixtures, or appliances at the other end. The head that is applied to the pipe and the fitting during the soldering operation causes a buildup of air pressure in the pipe system.

The built-up pressure, if it is not relieved by opening a faucet on the piping system, will cause the solder to be blown out of the soldered joint, leaving a hair-line crack that will then leak when the water is turned on.

Clean the male ends of the copper pipe and the female ends of the copper fitting with sandcloth or sandpaper. When clean, they will be bright and shiny. Then coat the cleaned ends with solder paste. Join the pieces together and apply heat from your torch. Unroll about 6″ of wire solder from the spool and apply the end of the solder to the joint. As soon as the solder starts to melt, apply the heat to the center of the fitting

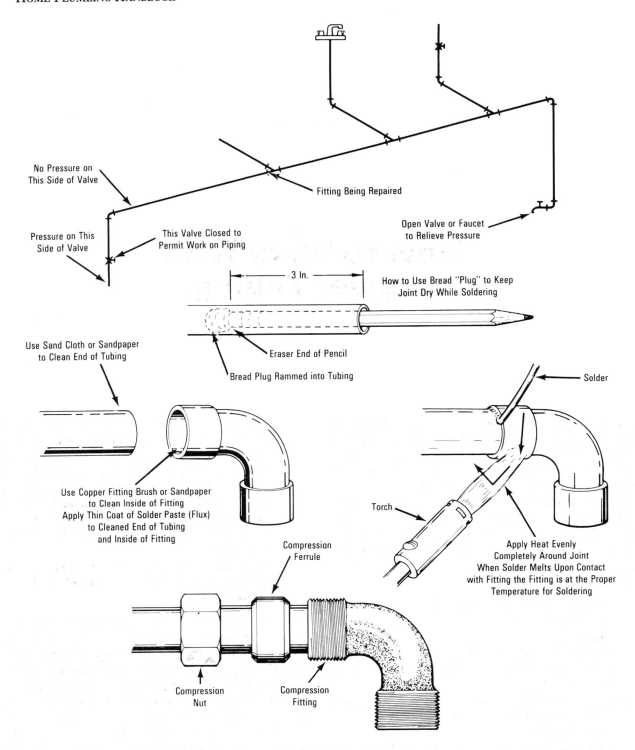

No Pressure on
This Side of Valve

Fitting Being Repaired

Pressure on This
Side of Valve

This Valve Closed to
Permit Work on Piping

Open Valve or Faucet
to Relieve Pressure

3 In.

How to Use Bread "Plug" to Keep
Joint Dry While Soldering

Use Sand Cloth or Sandpaper
to Clean End of Tubing

Eraser End of Pencil

Bread Plug Rammed into Tubing

Solder

Use Copper Fitting Brush or Sandpaper
to Clean Inside of Fitting
Apply Thin Coat of Solder Paste (Flux)
to Cleaned End of Tubing
and Inside of Fitting

Torch

Apply Heat Evenly
Completely Around Joint
When Solder Melts Upon Contact
with Fitting the Fitting is at the Proper
Temperature for Soldering

Compression
Ferrule

Compression
Nut

Compression
Fitting

Fig. 6-1. Making watertight connections with solder or compression joints.

and all the way around it. Follow the flame around with the solder. As soon as the solder has melted into the joint all the way around, remove the heat and allow the joint to cool. When all of the joints have been soldered, close the valves or faucets that were opened and turn the water back on. *Do not* put the tip of flame directly on joint to be soldered—the flame will contaminate the joint and the solder will not flow into the joint.

As I stated earlier, there is an easy way to make repairs or add fittings or piping to an existing piping system that requires no soldering. Ferrule-type com-

pression fittings are used. These fittings are readily available at plumbing shops and hardware stores. When using this type of fitting, insert tubing into the fitting recess and tighten compression nut onto fitting. Tightening the compression nut will lock the ferrule onto the tubing and if the compression nut is properly tightened, the joint will be water-tight. The ferrule-type compression fitting requires no soldering or solder paste, nor does it require cleaning of the tubing or the fittings.

There is still another way of installing fittings in copper piping. This last method requires the use of a flaring tool, or a flaring block. "Hard" copper cannot be flared unless it is annealed, or softened. Hard copper tubing can be annealed (softened) by heating it to a cherry-red color, then dipping it into cold water to cool it.

Flare-type fittings are designed for use on soft copper tubing, such as water service piping from the water main in the street into a building, or fuel oil piping from the oil tank to the furnace or boiler.

Pictured in Fig. 6-2 are some of the common solder-type copper fittings. Street fittings are any fittings with one male end and one or more female ends. A street tee would have one male end, two female ends. Copper street fittings have no restriction at the male ends and work very well in copper pipe installations. Tees and ells are available both as straight pipe size fittings and as reducing ells or tees. A tee is read (or described) in the following manner and as pictured: end 1, end 2, side 3. A straight ½" tee is ½" × ½" × ½". A reducing tee, such as ¾" × ½" × ¾" tee would be ¾" one end, ½" other end, ¾" on the side. The rule for describing a reducing tee is always (1) end size, (2) end size, (3) side size. Male and female adapters are used when connecting copper tubing to iron pipe. Copper tubing for plumbing use is measured by I.D. (Inside Diameter).

Copper tubing is measured and cut using the same rules as for steel piping. The most accurate method of measuring is the end-to-center measurement. The end-to-center measurement of a copper tubing fitting is very easy to figure because the fitting, whether it is a ferrule-type fitting or a solder-type fitting, is recessed for the piping. The distance from the end of the piping recess to the center of the fitting is the end-to-center

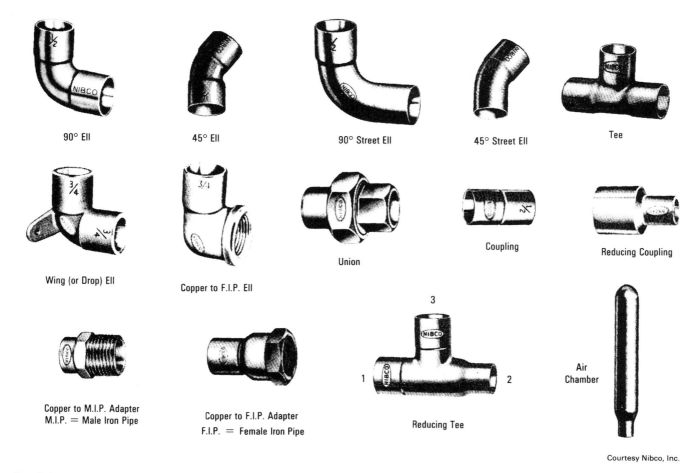

90° Ell   45° Ell   90° Street Ell   45° Street Ell   Tee

Wing (or Drop) Ell   Copper to F.I.P. Ell   Union   Coupling   Reducing Coupling

Copper to M.I.P. Adapter
M.I.P. = Male Iron Pipe

Copper to F.I.P. Adapter
F.I.P. = Female Iron Pipe

Reducing Tee

Air Chamber

Courtesy Nibco, Inc.

Fig. 6-2. Various solder connection fittings.

measurement of the fitting. The formula for figuring 45° breaks in the piping is the same as shown in the chapter on how to measure and cut steel piping (90° break × 1.41).

Copper-plated hangers, straps, or pipe hooks should always be used for hanging or securing copper piping. Copper-plated hangers prevent electrolysis or corrosion between dissimilar metals.

A soldered (sweat) joint can be made in any position. When heat is applied to a fitting, it should be heated away from the actual joint, toward the center of a tee or an elbow, for instance. Horizontal or vertical joints solder equally well; the heat "pulls" the solder up and into the joint.

Three types of copper are in common use. Type K is available in both hard and soft form. Type K copper is used primarily for water services or in buried or inaccessible locations. Type L is also in both hard and soft forms, and is used primarily for above-ground and inaccessible locations. Type M is a light-weight, thin-wall copper, available only in hard form, and is used primarily in open areas.

CHAPTER 7

How to Measure and Cut Pipe

There are many times in the course of a remodeling project, or when adding new plumbing fixtures or repairing a leak, when knowing how to measure the needed pipe can save time and work. Example: Fig. 7-1(A) shows tees added into existing piping and new piping added as connections for an automatic washer. In this case, we will assume the pipes are on the same level against the floor joists on the ceiling of a basement. Because the hot and cold piping is on the same level, it will be necessary to use fittings to enable the new pipe to pass under, or over, the existing pipe.

Fig. 7-1(B), (C), and (D) show different ways in which this can be done. In Fig. 7-1(B) and (C), the new piping is on the same level. In example (D), the new hot-water pipe would be above the cold-water pipe. In Fig. 7-1(B) and (C), the two pieces of vertical piping would be the same length. In Fig. 7-1(D), the new vertical hot-water pipe would be longer than the new cold-water pipe. If you have your own pipe dies and vise, the necessary cuts and threads can be made on the job. A small job, such as that illustrated, would not justify the cost of renting these tools. Most hardware stores that stock pipe and fittings will also thread pipe.

The correct and most accurate way to measure out lengths of pipe is to make an "end-to-center" measurement. Fig. 7-1(A) shows a section of piping with a tee, nipple, and union installed between points (3) and (4). "End-to-center" means from the end of the pipe,

with the fitting "made-up," or tightened onto the pipe, to the center of that fitting. In order to make an end-to-center pipe measurement, it is necessary to know how to measure the end-to-center distance of a fitting. Fig. 7-2(A) illustrates a full size drawing of a ½" elbow—X is the distance the thread makes up into the fitting, (½"). The distance Y is from the end of the thread of the pipe to the centerline of the fitting.

In Fig. 7-1(E) we show a 3-inch nipple "made-up" into a tee. The nipple length (3") plus the end-to-center measurement of the fitting (⅝") gives an end-to-center pipe measurement of 3⅝". Suppose you want to cut a piece of pipe to measure 24¾" end-to-center of a tee; the actual length of the pipe would be 24⅛". The 24⅛" plus the ⅝" (end-to-center of the fitting) would equal 24¾" end of the pipe to center of the tee, or the desired length. Also shown in Fig. 7-2 is a full size ½" tee, ½" 45° elbow, and a ½" union.

Referring again to Fig. 7-1(F) the view is an isometric drawing of the piping shown in Fig. 7-1(A).

From the front side, facing any fixture, the hot-water pipe should always be on the left side. The tees shown in Fig. 7-1(A) should be placed in such a manner to bring the piping to the desired location, with the hot-water pipe on the left. When a fitting is added to piping between other fittings (3) and (4), and (3) and (4) cannot be turned, a union must be used to join the new pipe. The distance between (5) and (6) is the end-to-end measurement of the present piping.

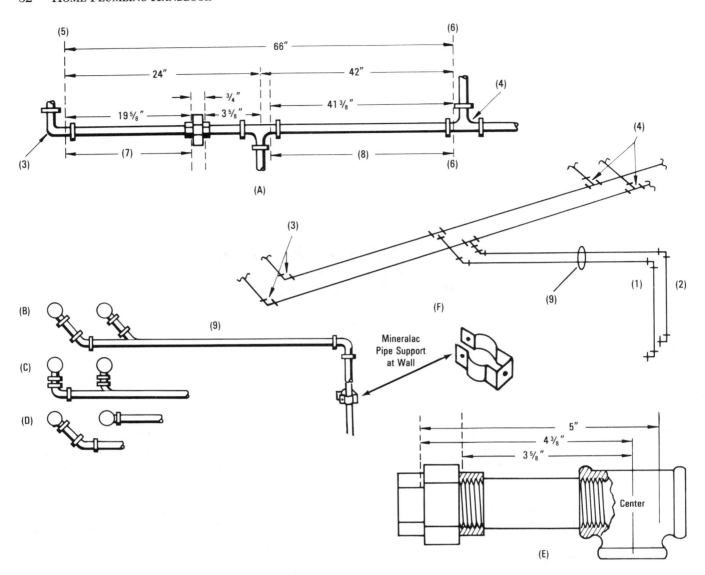

Fig. 7-1. Various ways of measuring pipe.

After a tee, a nipple, and a union are added, or cut into the present piping, this distance must remain the same; therefore, some pipe must be cut out to make room for the tee, union, and nipple.

Fig. 7-1(A) explains how the actual cut lengths of pipe are determined. For the purpose of this explanation, the original piping length between (5) and (6) is 66". If the center of the tee in the cold-water pipe is to be 24" from the makeup point (5), then the actual cut pipe length (7) would be 19⅝" (19⅝" + 4⅜" = 24"). The other cut piece [Fig. 7-1(B)] would then be 66" minus 19⅝" + 5" (or 24" end-to-center of tee plus ⅝" to make-up of tee) or 41⅜" end-to-end.

To determine the end-to-center measurement of (9) in Fig. 7-1(F) and Fig. 7-3, and using the type of pipe

support shown, the distance from the wall line to the center of the pipe support, when the support is mounted on the wall, is ¾". If a rule is held against the wall as shown and the end of the pipe (where the pipe makes up into the 45° elbow) to the wall measurement is 34", then the end-to-center measurement of this pipe would be 34" minus ¾" (to center of pipe support), or 33¼" end-to-center of the elbow. The drop pieces of pipe (1) and (2) can be cut to any desired lengths. All of the end-to-center measurements given in example are for ½" pipe; end-to-center measurements for other sizes of pipe and fittings are arrived at in exactly the same way. The table in Fig. 7-3 shows these measurements for other common pipe sizes.

Quite often it is necessary to offset around a light

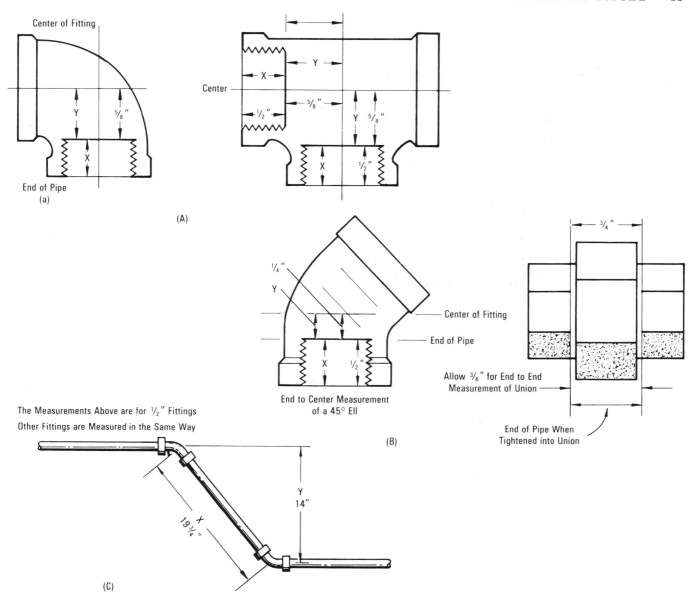

Center of Fitting

Y

$\frac{5}{8}''$

X

End of Pipe
(a)

(A)

Center

Y

X

$\frac{1}{2}''$

$\frac{5}{8}''$

Y

$\frac{5}{8}''$

X

$\frac{1}{2}''$

$\frac{1}{4}''$

Y

X

$\frac{1}{2}''$

Center of Fitting

End of Pipe

End to Center Measurement
of a 45° Ell

(B)

$\frac{3}{4}''$

Allow $\frac{3}{4}''$ for End to End
Measurement of Union

End of Pipe When
Tightened into Union

The Measurements Above are for $\frac{1}{2}''$ Fittings
Other Fittings are Measured in the Same Way

Y
14"

X
$19\frac{3}{4}''$

(C)

Fig. 7-2. How to measure fittings.

fixture, a post, or other obstructions when installing new runs of pipe. There is a formula for figuring the lengths of pipe when offsets are necessary. A 45° fitting should be used when making offsets because there is less friction loss (and therefore less pressure drop) through a 45° fitting than there is through a 90° fitting. For the purpose of this illustration, we will assume it is necessary to offset both the hot- and cold-water pipe 14" in order to go around a light fixture. The formula for this offset is the square break, 14" × 1.41. The result of this is the center-to-center distance between two 45° ells. Subtract the end-to-center measurements

of the fittings (2-45° elbows), and the answer is the length of the pipe. Fig. 7-2(C) shows how to figure the offset: Y is the distance of the square break; X is the center-to-center measurement of the 45° break; therefore, X = 14" × 1.41, or 19.74. We round off 0.74 to 0.75, and the answer 19.75 (19¾") is the center-to-center measurement of the 45° break. In Fig. 7-2(B), the end-to-center measurement of a ½" 45° elbow is ¼". There is a 45° elbow at each end of the 45° break, so we deduct ¼" × 2 = ½". Subtract ½" from the center-to-center measurement, and the answer 19¼" is the length of the pipe in the 45° break.

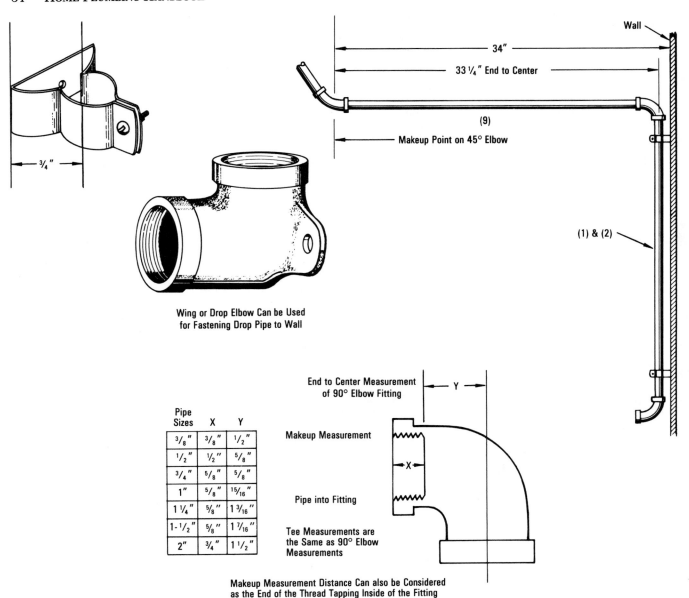

Pipe Sizes	X	Y
3/8"	3/8"	1/2"
1/2"	1/2"	5/8"
3/4"	5/8"	5/8"
1"	5/8"	15/16"
1 1/4"	5/8"	1 3/16"
1-1/2"	5/8"	1 7/16"
2"	3/4"	1 1/2"

Fig. 7-3. Makeup measurements.

When pipe cutting and threading is being done, care must be taken to ream the burr from the cut ends. This can be done with a file if a regular type of pipe reamer is not available. The new pipe thread being cut will be the right length when the new thread is flush with the outside face of the pipe die. A thread that is too long will be loose and prone to leak; a thread that is too short will be hard to start in a fitting.

CHAPTER 8

PVC and CPVC
Pipe and Fittings

The Age of Plastics is here, and no one can benefit from it more than the homeowner, the do-it-yourself-er, for whom this book was written. The hard work involved and the relatively high cost of the materials once used, together with the great number of different tools that were needed, meant that in the past many do-it-yourself projects were almost impossible. Those days are gone, thanks to PVC and CPVC pipe and fittings.

For years plumbing codes in many areas required the use of cast-iron soilpipe for underground soil (human waste) and drainage piping, and for above-ground main stacks, as shown in Fig. 1-1. These codes also required the use of galvanized steel pipe for branch waste and vent piping; in some areas lead pipe was required for waste connections to sinks, lavatories, and toilets.

Later, many of these areas permitted the use of DWV (drainage-waste-vent) grade copper tube for above-ground soil, waste, and vent piping. Economics—the rising cost of copper and the comparatively low cost of PVC (polyvinyl-chloride)—have brought about the extensive use of PVC pipe in the plumbing industry. Here are some of the reasons for using PVC pipe and fittings:

1. No special tools are needed for installing PVC. The only tools required are a rule to take measurements, a saw or cutter to cut the pipe, and a knife or file to ream out the burrs at the cuts.

2. The extremely smooth inside surface is corrosion-resistant, preventing the build-up of scale, rust, and foreign material that often impedes flow through metallic pipes.

3. PVC pipe and fittings are light in weight and are easily and quickly installed with chemically welded joints through solvent cementing.

4. PVC can be connected to existing cast-iron soil pipe, using either a poured lead joint or a threaded adapter.

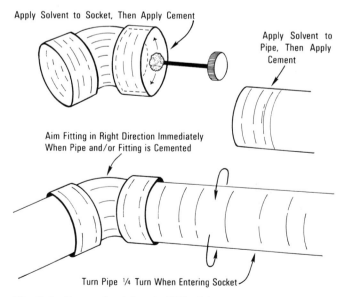

Apply Solvent to Socket, Then Apply Cement

Apply Solvent to Pipe, Then Apply Cement

Aim Fitting in Right Direction Immediately When Pipe and/or Fitting is Cemented

Turn Pipe ¼ Turn When Entering Socket

Fig. 8-1. Cementing joints in PVC piping.

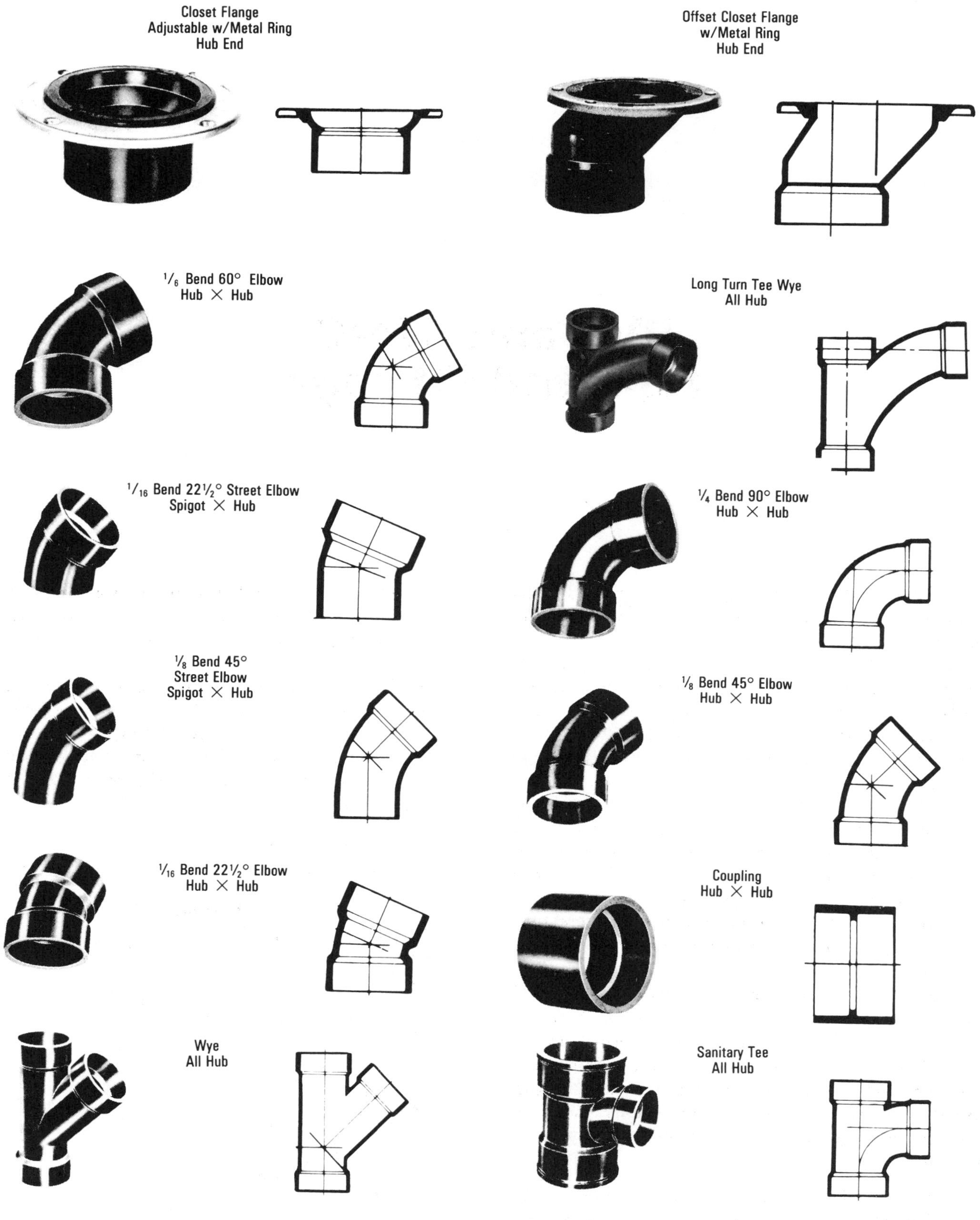

Closet Flange
Adjustable w/Metal Ring
Hub End

Offset Closet Flange
w/Metal Ring
Hub End

$\frac{1}{6}$ Bend 60° Elbow
Hub × Hub

Long Turn Tee Wye
All Hub

$\frac{1}{16}$ Bend 22$\frac{1}{2}$° Street Elbow
Spigot × Hub

$\frac{1}{4}$ Bend 90° Elbow
Hub × Hub

$\frac{1}{8}$ Bend 45°
Street Elbow
Spigot × Hub

$\frac{1}{8}$ Bend 45° Elbow
Hub × Hub

$\frac{1}{16}$ Bend 22$\frac{1}{2}$° Elbow
Hub × Hub

Coupling
Hub × Hub

Wye
All Hub

Sanitary Tee
All Hub

Courtesy Colonial Engineering, Inc.

Fig. 8-2. There is a PVC fitting to fit every need.

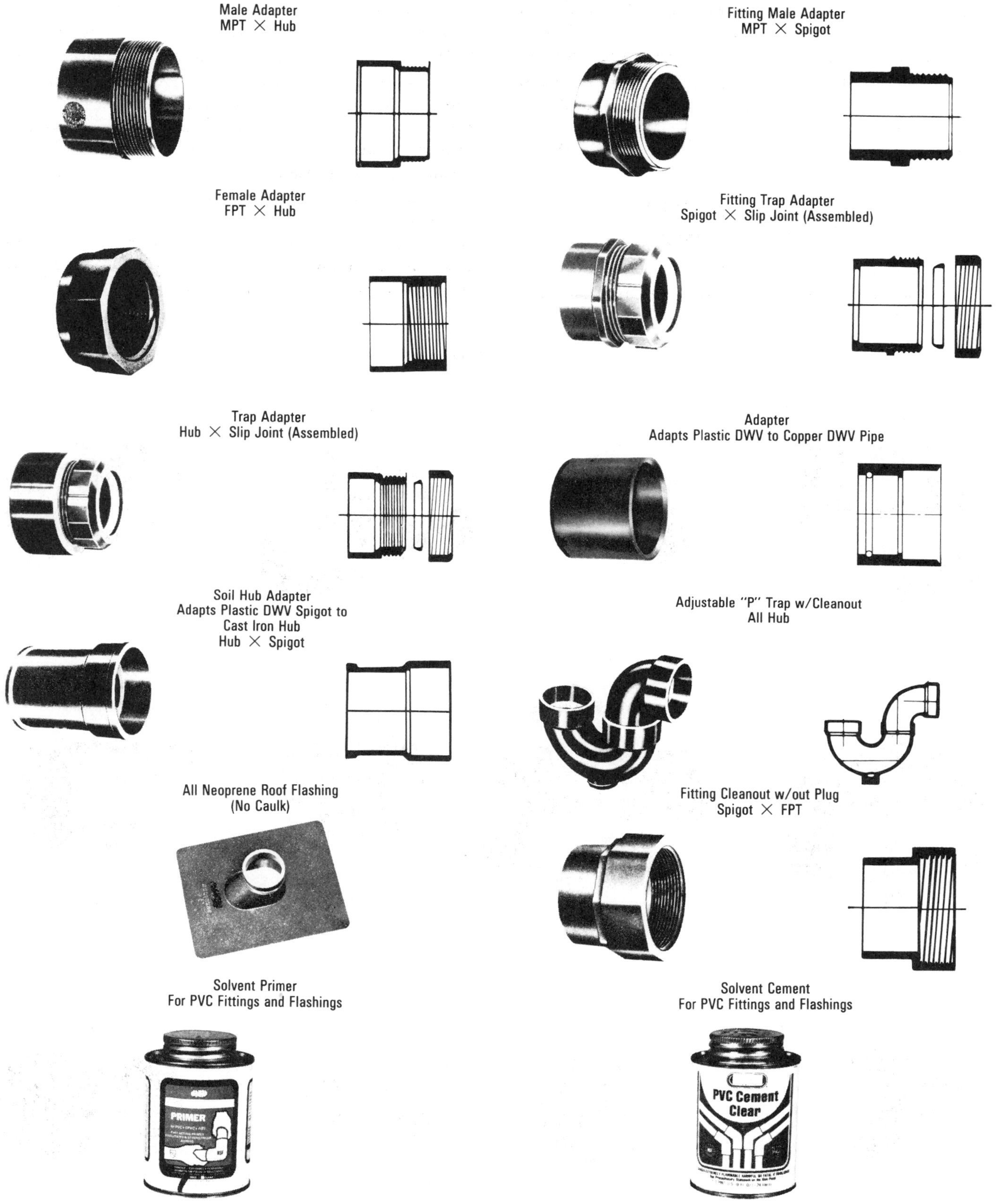

Male Adapter
MPT × Hub

Fitting Male Adapter
MPT × Spigot

Female Adapter
FPT × Hub

Fitting Trap Adapter
Spigot × Slip Joint (Assembled)

Trap Adapter
Hub × Slip Joint (Assembled)

Adapter
Adapts Plastic DWV to Copper DWV Pipe

Soil Hub Adapter
Adapts Plastic DWV Spigot to
Cast Iron Hub
Hub × Spigot

Adjustable "P" Trap w/Cleanout
All Hub

All Neoprene Roof Flashing
(No Caulk)

Fitting Cleanout w/out Plug
Spigot × FPT

Solvent Primer
For PVC Fittings and Flashings

Solvent Cement
For PVC Fittings and Flashings

PRIMER

PVC Cement
Clear

Courtesy Colonial Engineering, Inc.

Fig. 8-3. Knowing what to ask for makes the job easier.

5. PVC is virtually acid-proof to any chemical used in recommended strengths around the home. Certain chemicals such as methyl-ethyl-ketone, used in paint removers and paint brush cleaners, should not be poured in PVC or any other drainage pipe.

6. The best reason of all: you can do it yourself!

INSTALLING PVC PIPE AND FITTINGS

Remodeling projects, home improvement, and common plumbing repairs often require the installation of new drainage and water piping. PVC-DWV pipe and fittings are used for drainage and vent piping only. CPVC pipe and fittings are used for water piping, and the measuring and assembly of CPVC piping and fittings is done exactly the same as when using PVC-DWV pipe. CPVC pipe and fittings have very good insulating ratings, with comparatively small heat loss through hot-water piping and little or no condensation or sweating on cold-water piping.

The first step in installing the piping is to make accurate measurements. Chapter 7, **How to Measure and Cut Pipe**, explains end-to-end and end-to-center measurements and makeup points of fittings, etc. A fine-

CPVC Fittings

90° Elbow	45° Elbow	Tee	Coupling
Union	Reducing Coupling	Male Adapter	Female Adapter
Cap	Plug	Reducing Bushing	Nipples MPT

Courtesy Colonial Engineering, Inc.

Fig. 8-4. A few of the wide range of CPVC fittings.

toothed hand saw or a tubing cutter can be used to cut the pipe to length. If a saw is used, a miter box should also be used to make square cuts.

The outside of the pipe and the socket of the fitting should be wiped clean with a rag. In order to assure adequate penetration and fusion of solvent cement when joining PVC components, a primer (cleaner) must be used. The primer should be applied first to the inside socket surface, using a scrubbing motion. Repeated applications may be needed to soften this surface. Then apply a liberal portion of the primer to that portion of the male end of the pipe that will enter the fitting socket. Finally, apply one more coating of solvent to the fitting socket and then, while the socket is still wet, apply the cement to the socket and to the cleaned area of the pipe, and quickly insert the pipe into the socket. Turn the pipe a quarter turn while inserting it into the socket, and when it has bottomed out, hold it in place for 30 seconds to prevent the pipe from "backing out." When joining pipe and fittings, if the fitting must be turned to a certain direction or angle exactly, this must be done *immediately*, within 3 or 4 seconds. Applying the solvent and cement is shown in Fig. 8-1. After the cement has taken its initial set, the joint cannot be turned. Both the solvent and the cement usually have a dauber built into the top of the can; if they do not, a small, soft bristle brush can be used to apply the primer and the solvent.

When buying cement and solvent, use PVC cement and PVC solvent on PVC pipe and fittings. ABS-DWV fittings can be used with PVC-DWV pipe but when they are used together a universal-type cement and solvent, made for use with either PVC or ABS must be used.

Figs. 8-2 and 8-3 show some of the wide variety of PVC-DWV fittings and their names, to help you know what to ask for.

CPVC is being widely used for hot- and cold-water piping. CPVC is not recommended for hot-water usage where the temperature of the water is 210°F or above, but this is not a problem for the homeowner. In the home, water heaters should not be set to produce water temperatures above 120°F. Water temperatures above 120°F can cause severe burns.

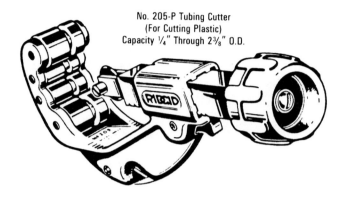

No. 205-P Tubing Cutter
(For Cutting Plastic)
Capacity ¼" Through 2⅜" O.D.

Courtesy Ridge Tool Co.

Fig. 8-5. Plastic tubing cutter.

CPVC piping is easily installed. No special tools are needed, and if installation instructions are followed, they will result in perfect leakproof joints. Like all other plumbing piping, both PVC and CPVC piping must be protected against freezing temperatures.

CHAPTER 9

Installing a
Garbage Disposer

Garbage disposers, once a luxury appliance, are now installed in almost all new homes. And like everything else we use or wear, after a certain time garbage disposers seem to self-destruct. But there is a way of getting your money's worth, either when buying a new home with a disposer installed or when buying a disposer for your present home. Manufacturers usually make several models that vary widely in price. If you stick to middle-of-the-line and up models, you'll be buying the most value for your money. These models will run quieter, last longer, and jam less than cheaper models. Other things to consider when buying a disposer are: ease of installation, availability of repair service and parts, and ease of unjamming. A ¼″ offset Allen wrench is furnished with *In-Sink-Erator* disposers. The wrench can be inserted into an opening in the bottom of the disposer and used to turn and free a jammed disposer. The disposer shown in Fig. 9-1 has an automatic reversing feature; each time it operates it rotates in the direction opposite to its previous rotation. This automatic reversing feature helps to prevent jamming.

Disposers are made in two types, continuous feed and batch type. Continuous feed types are operated by a remote switch, usually mounted on the wall above or at the side of the kitchen sink. They will accept garbage while running. Batch feed types are made with the electrical switch in the flange assembly. When this type is loaded, the lid is inserted in the flange and turned. This action turns the disposer on. It will operate until the lid is turned to OFF position and

Courtesy In-Sink-Erator, Emerson Electric Company

Fig. 9-1. A twist of the locking ring locks this disposer in place.

removed. This type of disposer cannot be refilled with garbage while it is running. It is almost impossible to turn a batch-type disposer on accidently—an obvious safety feature which is well worth considering. And remember, no matter which type of disposer is used, cold water, and plenty of it, should be running into the disposer when it is being used.

Good step-by-step instructions can make what would otherwise be a difficult job into a pleasant morning's work. So if you need a new disposer, read through the instructions that follow, decide which disposer to buy, and start to work. After reading the instructions, you may realize that the mounting procedure of the *In-Sink-Erator*, in which the mounting assembly is installed first and the disposer unit is then hung on the mounting assembly, is a big plus. The installer does not have to fight the weight of the unit while installing the mounting assembly. The disposer is attached to the mounting assembly by an easy twist of the lower mounting ring on the disposal unit.

Regardless of the brand of disposer being installed, don't forget to knock out the plug in the dishwasher connection before mounting the disposer if the dishwasher connection is to be used.

If this is your sink's first garbage disposer and an electrical outlet was not provided in the cabinet under the sink when the house was wired, this outlet must be provided now. A suggested method of wiring—connecting a length of 12/2 with ground, nonmetallic cable (*Romex*, etc.) to an existing receptacle above the sink, and extending it down inside the wall to a receptacle under the sink—is shown in Fig. 9-2, with a detailed illustration of wiring connections shown in Fig. 9-3. A three-wire plug and cord is connected to the disposer as shown in Fig. 9-2 and is then plugged into the receptacle under the sink.

The wiring shown in Fig. 9-3 requires that an additional box be added to the receptacle box to accommodate the switch for the garbage disposer. Metal junction boxes called sectional gang boxes are used for

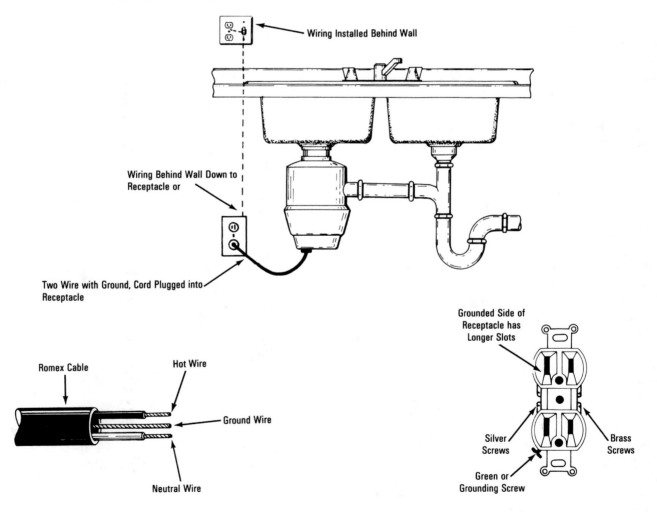

Fig. 9-2. Electrical wiring to disposer.

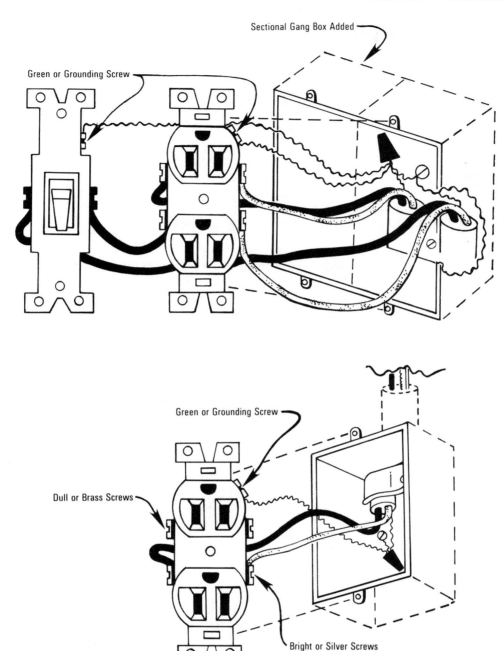

Sectional Gang Box Added

Green or Grounding Screw

Green or Grounding Screw

Dull or Brass Screws

Bright or Silver Screws

Fig. 9-3. One method of adding wiring for a garbage disposer.

this purpose. The right side of one box and the left side of another are removed, and the two boxes are joined. A single wall plate, with a switch opening and a receptacle opening, is used to cover the larger box.

If you are replacing an old disposer, the wiring will already be installed. Before starting to remove the old disposer, turn off the electrical power to the disposer circuit at the service panel (fuse box or circuit breaker box).

One of the best tests of a do-it-yourself-type book is that it presents the subject matter in such a way that the reader cannot go wrong, provided the instructions are followed carefully. And the instructions must lead the reader through a step-by-step process which, if followed, will result in a job well done. In this chapter I decided that instead of writing my own instructions, I would put myself in the position of the reader who needs to replace a garbage disposer and has never

INSTALLATION DIMENSIONS

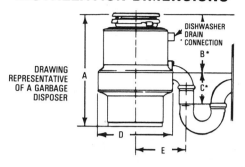

DRAWING REPRESENTATIVE OF A GARBAGE DISPOSER

MODEL #	A	B*	C*	D	E
CLASSIC	13¹¹/₁₆″	6¹³/₁₆″	4″	8⁹/₁₆″	7¹/₈″
CLASSIC LC	16¹/₁₆″	9⁷/₁₆″	4″	8⁹/₁₆″	7¹/₈″
77	13⁷/₁₆″	6¹³/₁₆″	4″	8¹/₂″	5³/₄″
17	15⁷/₈″	9⁷/₁₆″	4″	7¹/₂″	5³/₄″
333/SS	12³/₄″	6¹¹/₁₆″	4″	7³/₁₆″	5³/₄″
333	12³/₄″	6¹¹/₁₆″	4″	7″	5³/₄″
BADGER I	11³/₈″	5¹⁵/₁₆″	4″	6⁵/₁₆″	5″
BADGER V	12⁵/₈″	5¹⁵/₁₆″	4″	6⁵/₁₆″	5″
BADGER X	12⁵/₈″	5¹⁵/₁₆″	4″	8¹/₁₆″	5″

B*—Distance from bottom of sink to center line of disposer outlet. Add ¹/₂″ when stainless steel sinks are used.

C*—Length of discharge tube from center line of disposer outlet to end of discharge tube.

IMPORTANT: Plumb waste line to prevent standing water in disposer motor housing.

1 START BY CLEANING YOUR SINK'S DRAIN LINE. (If you have a brand new home, you can skip this step.)

The cutting elements on your old disposer were probably worn and not grinding the waste completely. Your drain line may be partially blocked. We recommend cleaning the line before connecting your new disposer.

You can do the job yourself with a drain auger. Remove the drain trap and, using the auger, clean out the horizontal drain pipe that runs from the trap to the main waste pipe.

2 ELECTRICAL SUPPLY

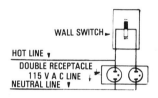

WALL SWITCH

HOT LINE
DOUBLE RECEPTACLE
115 V A C LINE
NEUTRAL LINE

In the future you may want to install an I-S-E hot water dispenser. It comes with a convenient 3 wire plug and cord. To accommodate it you may want to install a double wall receptacle for use with both your disposer and future hot water dispenser. During installation, one receptacle should be installed with a wall switch in the circuit, the other should be wired direct like a standard wall outlet.

Before you attack this job, you should be thoroughly familiar with electrical power and proper procedures. If you aren't, call in a professional who is knowledgeable.

This appliance is equipped with copper wires. Use 3 conductor copper cable in accordance with your local code to make your connections to the unit.

NOTE: See 10 & 11 for final electrical connections.

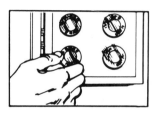

First, remove the fuse to open the circuit breaker on the circuit you plan to use for your disposer. Use a separate 15 amp, 115 volt circuit just for the disposer. If you are replacing a unit, skip to step 3. Next, use 15-20 amp, 115 volt cable to make a connection from the junction box to the on-off switch.

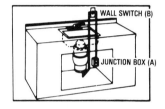

WALL SWITCH (B)

JUNCTION BOX (A)

For Continuous Feed Models 77, Classic 333/SS, 333 and Badger Models Only: Install a junction box and switch (Obtain ¹/₂ H.P. 15-20 amp rated on-off switch and electrical wiring locally.) Position the switch in any convenient location, and connect to junction box. *All wiring must comply with local electrical codes.* 14 gauge size wire is the smallest permissible for use with a 15 amp circuit, and 12 gauge size wire is the smallest permissible for use with a 20 amp circuit.

Important note for Models 17 and Classic/LC Batch Feed only. No separate wall switch (A) is needed; switch is built into unit. Go right to 'Electrical Connections'.

All wiring must comply with local electrical codes.

3 CHECK THE PARTS AGAINST THE DRAWINGS BELOW AND MAKE SURE EVERYTHING IS THERE. (Your new disposer may not look like the one pictured.)

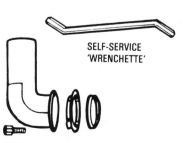

SELF-SERVICE 'WRENCHETTE'

DISCHARGE TUBE, GASKET, METAL FLANGE AND SCREW(S)

THE MOUNTING ASSEMBLY CONSISTING OF:

Sink flange

Fiber gasket

Back-up ring

Mounting ring and 3 screws

Snap ring

Mounting gasket

Lower mounting ring

THE DISPOSER ITSELF

Courtesy In-Sink-Erator, Emerson Electric Co.

4

HERE IS WHAT YOU DO IF YOU ARE REPLACING AN OLD DISPOSER.

FIRST, TURN OFF ELECTRICAL POWER at the service panel (fuse box or circuit breaker box). If the mounting is the same as your new disposer's mounting, you can use the existing mounting. Follow instructions A through E, Step 4, then go on to Step 10.

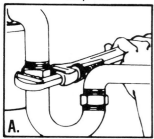

A.

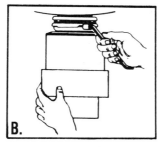

B.

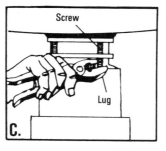

C.

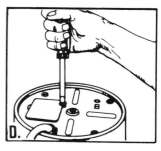

D.

Use a pipe wrench to disconnect the drain line where it attaches to the disposer discharge tube.

IF YOUR OLD DISPOSER HAS A DIFFERENT MOUNTING THAN YOUR NEW ONE, GO ON TO INSTRUCTION C.

If your old disposer has the same mounting as your new one, insert the end of your 'wrenchette' or screwdriver into the right side of one of the disposer mounting ring lugs at the top of the disposer. Then, turn the 'wrenchette' or screwdriver to the left (counterclockwise) until the lug lines up with one of the sink mounting assembly screws.

CAUTION: Be sure to hold the disposer with one hand while performing this step or it may fall when the mounting ring is disconnected from the sink mounting assembly. GO TO INSTRUCTION D.

If your old disposer has a different mounting than your new one, use a pliers or adjustable wrench to remove the nuts on the mounting ring. Then remove old disposer. (Some disposers have to be removed by taking off a clamp or by twisting the disposer to remove it from its mounting. Easy to figure out.)

Once the disposer is off, turn it upside down and remove the electrical plate.

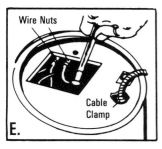

E.

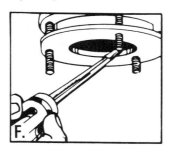

F.

G.

Now, use a screwdriver to remove ground wire. Remove the wire nuts from the power wires. Separate the disposer power wires from the cable wires. Loosen the screw(s) on the cable clamp and remove the cable from the disposer.

If your old disposer has a different mounting than your new one, follow steps F and G. Otherwise go on to Step 3.

Loosen screws and remove old mounting ring and back-up ring. You may need a hammer to loosen assembly parts.

Finally, remove the old sink sleeve by pushing it up through the sink hole.

5

HERE IS WHAT YOU DO IF YOU ARE INSTALLING YOUR SINK'S FIRST DISPOSER.

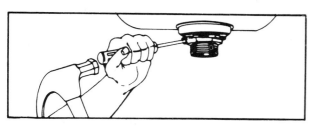

Now, remove the large-diameter nut at the base of the strainer by placing the tip of your screwdriver on the edge of the nut. (There are usually ridges to hold your screwdriver.)

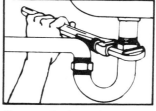

Use a wrench to loosen the nut at the top of the 'P'-trap.

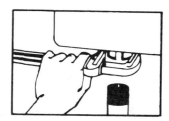

Next, remove the nut at the top of the sink strainer and remove the extension pipe.

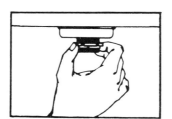

The nut should be loosened enough that you can spin it off by hand.

Now, push the strainer assembly up through the sink hole and remove it.

Courtesy In-Sink-Erator, Emerson Electric Co.

IF YOU ARE REPLACING AN OLD DISPOSER, CLEAN THE OLD SEALANT FROM THE RIM OF THE SINK HOLE.

Use your screwdriver or a putty knife to scrape away all traces of the old putty or caulking from the edge of the sink drain hole. Make sure you get this as clean as possible so that you'll have a good, watertight seal for your new disposer sink sleeve.

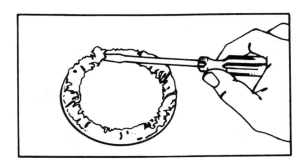

 NOW, WE'RE GETTING INTO THE ACTUAL INSTALLATION. SEPARATE THE PARTS IN THE MOUNTING ASSEMBLY.

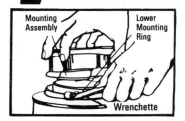

Start by disengaging the top mounting ring from the lower mounting ring. Do this by holding the top of the assembly with one hand and turning it counterclockwise while holding one of the mounting lugs with your other hand.

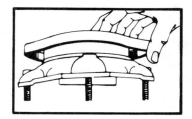

Then, loosen the screws on the mounting assembly until they are just level with the surface of the mounting ring.

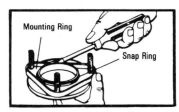

Now, use a screwdriver to pry off the snap ring.

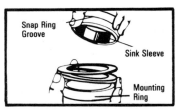

The assembly will now come apart. Set it aside and move to the next step.

8 APPLY PUTTY TO THE SINK FLANGE.

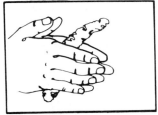

Buy ¼ lb. of non-hardening "plumber's" putty at your hardware store. Make a nice fat snake of putty by rolling it between your hands.

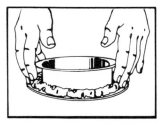

Apply this roll under the rim of the sink flange.

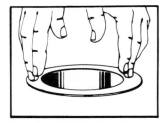

Then, place the sink's flange into the sink drain hole and push down gently but firmly to make sure it sits evenly in the putty.

Courtesy In-Sink-Erator, Emerson Electric Co.

NEXT, LET'S ATTACH THE UPPER MOUNTING ASSEMBLY.

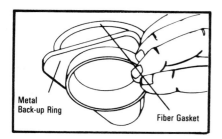

First, working from under the sink, slip the **fiber gasket** and next the **metal back-up ring** (flat side up) up and over the sink flange.

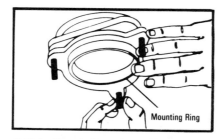

Hold the fiber gasket and metal back-up ring in place with one hand and place the mounting ring with its three screws *onto the sink sleeve.*

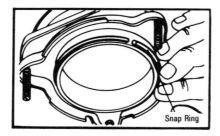

Now, push the fiber gasket, metal back-up ring and the mounting ring further up on the sink sleeve. Slide the snap ring onto the sink sleeve until it pops into place in the groove on the sleeve.

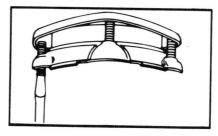

Tighten the three mounting screws with your screwdriver until the whole mounting assembly is seated evenly and tightly against the sink.

WE'RE READY TO MAKE THE ELECTRICAL CONNECTIONS.

Remove the electrical plate from the bottom of the disposer and pull out the two electrical wires. The ground screw is under the plate.

Connect the wires from the switch to the disposer wires. In making this connection, electrical wire nuts can be used or wires may be securely soldered together. *Be sure to connect white to white and black to black.* Wrap wire connections with electrical tape and put wires inside of disposer housing. Check for proper grounding, then replace the electrical cover.

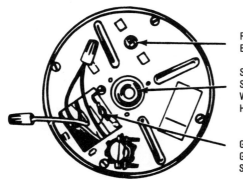

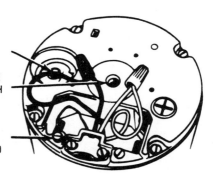

RESET BUTTON

SELF-SERVICE WRENCH HOLE

GREEN GROUND SCREW

Courtesy In-Sink-Erator, Emerson Electric Co.

11 NEXT, MAKE SURE THE DISPOSER IS GROUNDED.

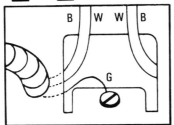

Simply attach the third green wire to the green ground screw on the unit.

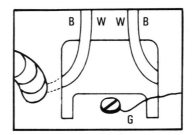

If you don't have a ground supply conductor, buy a length of copper wire from your hardware store that is no smaller in size than the supply wire and attach one end to the green ground screw on the unit.

WARNING:IF NOT PROPERLY GROUNDED, a hazard of electrical shock may exist. DO NOT reconnect electrical current at main service panel until proper ground is installed. For your safety, DO NOT ground to a gas supply pipe.

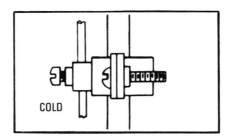

Attach the other end of this ground wire to the METAL cold water pipe. NOTE: Be sure that this cold water pipe is continuous METAL pipe from the sink to the ground. Use Underwriters Laboratories Inc. listed ground clamp to attach wire to pipe. If any NON-METAL pipe is used in your home water connections or if plastic pipe is used in your water supply pipe, you will need a qualified electrician to install a proper ground.

If you have a water meter in your home, check the meter to see if there is a wire that comes across it. If there is no wire, your cold water pipe is NOT GROUNDED. To properly ground it, add a #6 copper wire as shown for 200 amp service or less. Use Underwriters Laboratories Inc., listed ground clamps to attach wire to pipe.

12 PREPARING THE DISHWASHER DRAIN CONNECTION.
(If you have a dishwasher.) If you no *NOT* plan to connect a dishwasher drain to the disposer, go on to step 13.

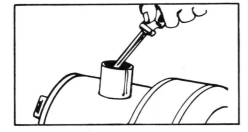

Lay the disposer on its side and insert the tip of your screwdriver into the dishwasher drain hole opening at an angle.

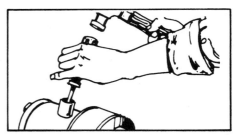

Rap the end of the screwdriver with a hammer until the molded plug pops out. Make sure you take the loose plug out of your disposer.

Courtesy In-Sink-Erator, Emerson Electric Co.

13 WE'RE ALMOST THERE. CONNECT THE DISPOSER TO ITS MOUNTING ASSEMBLY.

Lift the disposer and position it so that the disposer's three mounting ears are lined up **under** the ends of the sink mounting assembly screws.

Then, while holding the disposer in place, turn the lower mounting ring with the ears to the right until **all three** ears are engaged in the mounting assembly.

The disposer will now hang by itself. You will lock this ring later, after the plumbing connections are made.

14 NOW, ATTACH THE DISPOSER DISCHARGE TUBE TO YOUR SINK'S DRAIN TRAP.

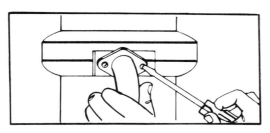

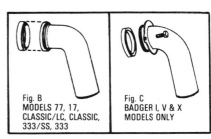

Fig. B
MODELS 77, 17,
CLASSIC/LC, CLASSIC,
333/SS, 333

Fig. C
BADGER I, V & X
MODELS ONLY

First, check inside the disposer grinding chamber to remove any foreign material that might have dropped in.

Rotate the disposer around and attach the discharge tube to the disposer. First, insert the rubber washer in the discharge opening. Then, put the metal flange over the discharge tube and screw the tube into place using the bolt(s) provided.

In fig. B, install the discharge tube gasket onto the discharge tube. Gasket must be installed as shown to assure a leak-proof installation. In fig. C, install discharge tube gasket into disposer discharge outlet. Gasket will be held in place by the discharge tube flange.

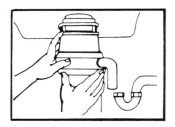

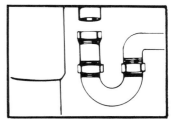

Rotate the disposer so that the discharge tube aligns with your drain trap.

TUBE TOO LONG? Simply cut off as much as necessary with a hack saw, making sure you have a clean, straight cut.

TUBE TOO SHORT? If the discharge tube doesn't reach your drain trap outlet, measure the shortness and buy a drain trap extension that includes a slip nut and install it in place.

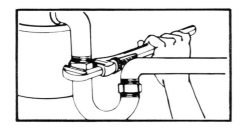

IF IT FITS—Simply tighten the slip nut on the trap to make your connection to the discharge tube complete. **IF DOUBLE SINK EXISTS, WE RECOMMEND USE OF SEPARATE TRAPS FOR DISPOSER AND SECOND SINK.**

NOTE: Be sure to comply with all applicable plumbing codes.

Courtesy In-Sink-Erator, Emerson Electric Co.

MAKE THE DISHWASHER DRAIN CONNECTION.

Make all connections to comply with local plumbing codes. An approved dishwasher connector kit, Part No. 6429D, is available from nearby I-S-E authorized service center. Use worm gear hose clamp on dishwasher connection.

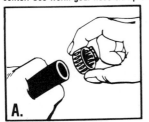

A.
Remove the clamp or fittings from the end of your dishwasher drain hose.

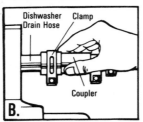

B.
Slide large end of rubber coupler, from dishwasher drain connection kit, over the inlet tube of the disposer. Fasten the coupler to disposer with the clamp provided.

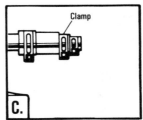

C.
Insert one end of the plastic tube into the coupler and fasten with $\frac{7}{8}$" clamp.

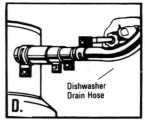

D.
Slip the remaining clamp (the one you purchased) over the dishwasher drain hose and back two or three inches. Now, slip the drain hose over the plastic tube, slide the clamp into place and tighten.

Note: Check the three clamps to be sure you have tightened all of them.

NOW, LOCK DISPOSER IN PLACE.   CHECK FOR LEAKS.

Place the end of your 'wrenchette' or a screwdriver into the left side of one of the disposer mounting lugs...at the top of the disposer. Then, turn the screwdriver or 'wrenchette' to the right until the disposer is firmly secured in position, engaging the locking notch.

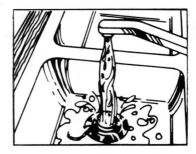

Run water slowly through the unit. Then, place the stopper in seal position and fill the sink with water.

Finally, remove the stopper and permit the water to flow. Check for leaks and correct, if necessary.

Courtesy In-Sink-Erator, Emerson Electric Co.

replaced one before. Therefore I tried to find the best installation instructions available, and settled on *In-Sink-Erator's* as the most detailed and clear.

Most do-it-yourselfers are very capable when it comes to many types of repair and replacement work around the home, but are hesitant to become involved with electric wiring. This is a good thing; if you don't feel confident when working with electricity, do the other parts of the job that you are confident with and leave the wiring to the experts. You'll still come out way ahead, and who knows, maybe while you are watching, some of the experts' knowledge will rub off!

CHAPTER 10

Replacing a Dishwasher

Because dishwashers, like all other appliances, wear out, sooner or later they must be replaced. Replacing a dishwasher is ordinarily much easier than a new installation because unless the dishwasher is very old, it is simply a matter of removing the old one and, using the same connections, connecting the new one.

First, shut off the hot-water supply to the old unit. Then turn off the electrical circuit to the dishwasher. If you are in doubt as to which fuse or breaker controls this circuit, remove the main fuses or switch off the main breaker and leave the electrical power off until the new dishwasher has been set and connected.

The top front of a dishwasher is usually secured to the bottom of the counter top by two screws that will need to be removed. There should be four leveling screws under the unit, one at each corner. These screws must be screwed in (up), dropping the dishwasher to a point where it can be pulled out. Disconnect the electrical wiring at the terminal on the dishwasher, then disconnect the water supply at the dishwasher and push it out of the way so that the dishwasher can be pulled out. An old towel or rug, placed under the old unit as it is pulled out, will make removal easier and will also prevent damage to the kitchen floor. It may be possible to reuse the old water supply piping, with some adjustment. If new supply piping is needed, it may be easier to install it to the approximate point of connection to the new unit, before setting the new one in place. Type L soft copper tubing is very easy to bend and shape into place.

Many manufacturers recommend *not less* than ⅜″ I.D. hot-water supply to their units. Copper tubing of ⅜″ I.D. measures ½″ O.D. and is much harder to bend without kinking than ⅜″ O.D. tubing. I have been connecting dishwashers for many years using ⅜″ O.D. tubing and have never had any trouble with the water supply to the units. If, following my example, you choose to make the job easier, where ⅜″ I.D. tubing and fittings are mentioned later on, substitute ⅜″ O.D.

If the drain hose is not connected to the dishwasher when you receive it, it should be connected before sliding the unit back into place. Start the drain hose through the hole below the counter top before sliding the unit into place. As you push the unit back, have someone pulling on the hose at the same time.

When the dishwasher is in place, the leveling screws should be screwed out (down) until the top front frame of the dishwasher is against the bottom of the counter top. Check each side of the dishwasher for clearance by opening and closing the door. Also check that the unit is lined up correctly with the front of the cabinets on each side. When you are satisfied with the placing and leveling of the unit, secure the top front to the bottom of the cabinet top.

Most manufacturers of electric dishwashers specify that the hot-water supply pipe to the dishwasher must not be less than ⅜″ I.D. Soft copper tubing is the easiest to install for this purpose. The pipe connection at the dishwasher inlet valve should be ⅜″ I.P.S. (iron

pipe size). An adapter fitting, either elbow type or straight type, depending on the make of dishwasher, should be used at this point.

If ⅜″ I.D. tubing is used to connect the hot-water supply to the dishwasher, an adapter with a ⅜″ M.I.P. (male iron pipe) thread on one end and a ⅜″ I.D.

tubing ferrule connection on the other end should be used. If ⅜″ O.D. tubing is used, the adapter should have a ⅜″ M.I.P. thread on one end and a ⅜″ O.D. tubing ferrule connection on the other.

The tee connection for the hot-water supply to the dishwasher is shown in fig. 10-1. Instructions for the

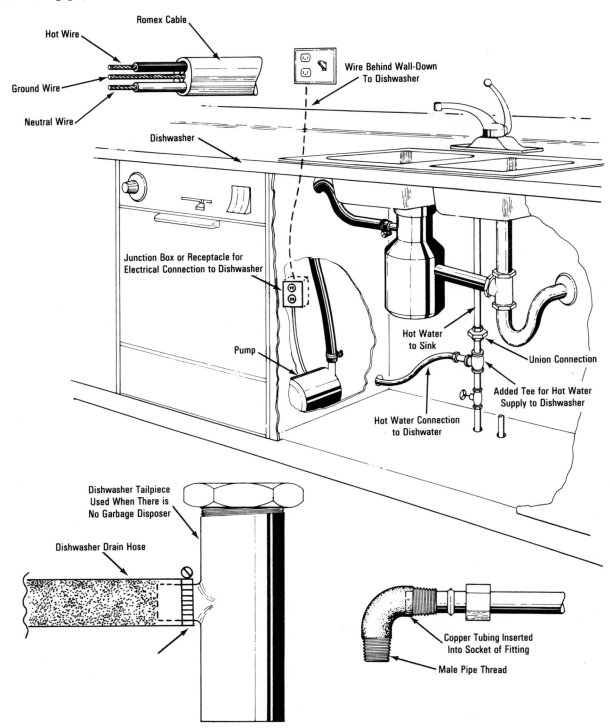

Fig. 10-1. Connections for automatic dishwasher.

electrical connections to the dishwasher will be found in the installation instructions furnished with the dishwasher.

If you have a garbage disposer on the sink, the dishwasher drain can be connected to the disposer. Garbage disposers are made with a separate opening on the side for dishwasher drains. Dishwashers are furnished with a drain hose, which connects to the disposer inlet with a rubber dishwasher connector. This connection is shown in Fig. 10-1. The dishwasher drain hose must be brought up to the bottom side of the counter top, then run through a hole in the cabinet and back down to the dishwasher drain inlet on the disposer, as shown in Fig. 10-1. The drain hose must be brought out through the side of the cabinet as shown to prevent water running out of the dishwasher while it is cycling, and also to prevent the wash water from siphoning out of the dishwasher. Some codes require a siphon breaker on the discharge piping between the dishwasher and the garbage disposer or dishwasher tailpiece.

Fig. 10-2 shows a 3" length of ⅝" O.D. copper tubing inserted into the end of the drain hose, which in turn is inserted into the end of the rubber dishwasher connector. A clamp is used at this point to make a watertight connection. The copper tubing provides a solid base to clamp to, and it will not impede water flow through the hose. A clamp is also needed at the connection to the disposer inlet. It may be necessary to cut off the first or smallest section of the rubber connector to allow the drain hose to enter.

If this is a new installation and the dishwasher drain connection on the disposer has not been used, it will be necessary to remove the plug at the inside edge of the disposer drain inlet. On some disposers this is a rubber plug, on others it is metal. This plug must be knocked out from the outside, and it can usually be removed by inserting a 5" or 6" length of ⅝" copper tubing into the inlet and rapping it sharply with a hammer.

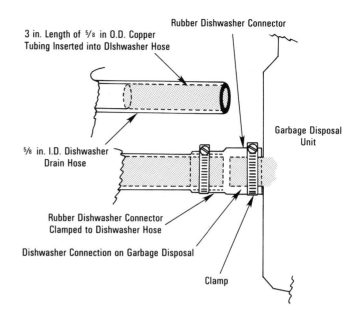

Fig. 10-2. Connecting the dishwasher to a garbage disposal unit.

One type of dishwasher tailpiece is shown in Fig. 10-1. If the dishwasher drain is to be connected to a tailpiece instead of a disposer, the connection could be as shown, and the hose slipped over the inlet on the tailpiece and clamped.

One layer of Teflon tape dope should be used on any threaded water or drainage connections. The tape should be started at the male end of the fitting and wrapped in the direction of the thread.

After the dishwasher has been set and all connections are made, open the water supply valve and check the piping for leaks. If there are none, the electrical power should be turned on and the water and drainage connections checked once again for leaks. The final step is to install the lower panels or trim strips that were left off for connection purposes.

All plumbing and electrical work must comply with local codes and ordinances.

CHAPTER 11

Repair and Replacement
of Water Heaters

Gas-fired heaters—Service or installation of a gas-fired water heater requires ability equivalent to that of a licensed tradesman in the field. Plumbing, gas supply, air supply, and venting are required.

Electric water heaters—Service or installation of an electric water heater requires ability equivalent to that of a licensed tradesman in the field. Plumbing and electrical work are required.

The above paragraphs are taken from the installation and service instructions furnished with a popular brand of water heaters. The following information is presented as a guide to trouble-shooting problems that arise with both gas and electric water heaters. Following the preventive maintenance procedures can avoid costly repairs or replacements.

If a heater tank is leaking, the heater must be replaced. There is no satisfactory repair for a leaking tank. If a relief valve is leaking, working the test lever several times in quick succession to wash out scale, etc., under the valve may stop the leak, otherwise the relief valve must be replaced.

GAS-FIRED WATER HEATERS

Following the correct pilot-lighting procedure will help in diagnosing problems with gas-fired water heaters. The procedure should be on a decal on the heater, or you can use the following:

Open the outer jacket door and remove the inner door to the burner chamber.

1. Turn the control knob (PILOT-ON-OFF) located on top of the thermostat to OFF position.
2. Wait five minutes—longer, if necessary—for any accumulated gas to escape.
3. Depress the reset button and turn the control knob to PILOT position.
4. Depress and hold down the reset button, light the pilot burner, and observe the thermocouple tip. It should be in the pilot flame; the tip should be heated a cherry red.
5. Hold the reset button down for at least one minute, then release it. If the pilot flame stays lit, the thermocouple is good. If the pilot flame goes out, check the thermocouple connection at the thermostat and tighten the connection if necessary. Repeat steps 1 through 5. If the pilot flame will not stay lit, the thermocouple should be replaced.
6. If the pilot flame stays lit, replace the inner door and the jacket door. Turn the control knob to ON and set the temperature dial to desired water temperature.

If the pilot flame stays on but the main burner will

not burn when the temperature dial is set to the desired water temperature, either the water is already at that temperature or there is a problem with the thermostat, possibly ECO operation.

Energy Cut-Off (ECO) Devices

ECO's are used on water heaters as an additional safety device. If the water temperature in the tank rises above normal operating temperature, the ECO is designed to operate (or open) and cut off the energy supply to the heater. An ECO is incorporated in the thermostat; there are two types, either of which may be used. One type is self-resetting and will reset when the hot water in the tank is replaced by cool water. The other type is a one-time ECO which, when it has operated, requires the replacement of the thermostat. A decal on the thermostat should identify which type is used. An ECO rarely operates due to thermostat failure, but may operate due to short draws. Short draws are short periods of hot-water usage repeated frequently over a short time period. This causes the thermostat to come on every few minutes, resulting in a build-up of very hot water that may trigger the ECO. When the ECO operates under these conditions the thermostat is not at fault, although if the thermostat has a one-time ECO, the thermostat must be replaced. If an ECO operates due to short draws, the usage pattern must be changed.

Relief Valve

The relief valve is a safety device installed on a water heater to prevent injury or damage that could occur due to the excessive temperature and pressure build-up caused by thermostat failure. A relief valve used on a water heater should be an ASME rated valve with a temperature setting of 210°F and a pressure rating of *not more than 150 psi* (pounds per square inch). The discharge piping from a relief valve must extend full size to within 6″ of the floor, or otherwise as local codes require, as shown in Fig. 11-3.

Thermocouple

The action of a thermocouple is to generate electricity by thermal action. This voltage, 7 millivolts or more, energizes a magnet in the thermostat. When the magnet is energized it permits gas to flow from the thermostat to the main and pilot burners. A thermocouple which when heated will not generate the necessary voltage to energize the thermostat magnet must be replaced. The thermocouple tip must be positioned correctly, as shown in Fig. 11-1, for thermal generation to take place.

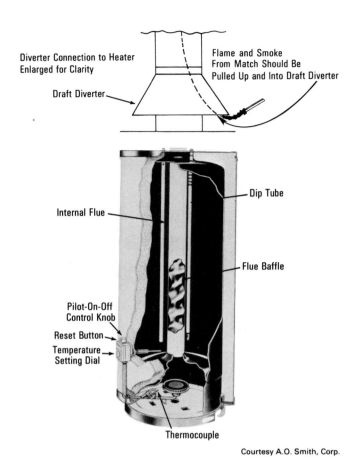

Courtesy A.O. Smith, Corp.

Fig. 11-1. Cutaway view of A.O. Smith gas heater.

The Trouble-Shooting Chart can help you determine what the trouble is with a malfunctioning gas water heater.

ELECTRIC WATER HEATERS

As mentioned earlier, the repair or installation of an electric water heater requires a high level of experience in both the plumbing and electrical fields. The following information is presented only to acquaint the homeowner with the problems that can occur with an electric water heater. A leaking water heater must be replaced; there is no satisfactory permanent repair. Heating elements and thermostats can be replaced by skilled personnel, saving the cost of a new heater when only these parts are needed. Testing for a suspected burned-out (open) element should only be done by a skilled technician, since a voltage of 240 volts (±10%) may be present.

Elements

Most household electric water heaters have two electric elements; lowboy types may only have one.

Troubleshooting Gas Water Heaters

Nature of Trouble	Possible Cause	Service Procedure
Pilot will not light	1. PILOT-ON-OFF knob not correctly positioned 2. Pilot orifice clogged 3. Pilot tube pinched or clogged 4. Air in gas line	1. Turn to pilot position, depress button fully, light pilot. 2. Clean or replace. 3. Clean, repair, or replace. 4. Purge air from gas line.
Pilot does not remain burning when reset button is released	1. Loose thermocouple 2. Defective thermocouple 3. Defective magnet in thermostat 4. Thermostat's one-time ECO has opened	1. Tighten connection at thermostat. 2. Replace thermocouple. 3. Call service technician. 4. Thermostat must be replaced.
Not enough hot water	1. Heater is undersized 2. Low gas pressure	1. Reduce rate of hot-water use. 2. Check gas supply pressure and manifold pressure.
Water too hot or not hot enough	1. Thermostat dial setting too high or too low 2. Thermostat out of calibration 3. High water temperature followed by pilot outage	1. Adjust dial setting as required. 2. Thermostat must be replaced. 3. Thermostat out of calibration; replace thermostat.
Yellow flame Sooting	1. Scale on top of burner 2. Combustion air inlets or flue ways restricted 3. Not enough combustion or ventilation air supplied to room 4. Insufficient draft in vent	1. Shut off heater and remove scale. 2. Remove lint or debris and inspect air inlet opening for restriction. 3. Call service technician immediately; turn off gas supply to heater. 4. Call service technician immediately; turn off gas supply to heater.
Rumbling noise	1. Scale or sediment in tank	1. Open drain valve, drain 3 or 4 gal. monthly.

Caution: For your safety, DO NOT attempt repair of thermostat, burners, or gas piping. Refer repairs to qualified service personnel.

Upper elements rarely burn out, while lower elements burn out primarily because mineral sediments, lime, etc., settle in the bottom of the tank and eventually cover the heating element completely. When a heating element is energized while not submerged in water, it will burn out. When mineral deposits have caused element burn-out, before the element can be replaced the mineral deposits must be removed. To prevent these deposits from accumulating, three or four gallons of water should be drained from the heater every month, through the drain valve at the bottom of the tank.

Adjusting Thermostat Setting

If water temperatures are too hot or too cold, the thermostat can be adjusted to desired setting. The thermostat(s) are located under an access panel on the jacket of the heater. The thermostat, except for the adjusting dial, should be protected by a fiber cover to prevent contact with electrical wiring or connections. When adjusting the thermostat, *do not remove this cover*. It protects you from contact with high voltage.

Maintenance, Health, and Safety

Sediment that collects at the bottom of a heater tank is formed during water heater operation. This sediment should be drained every month or so, particu-

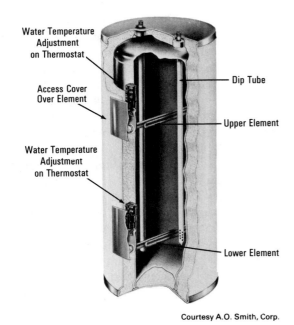

Water Temperature Adjustment on Thermostat

Access Cover Over Element

Water Temperature Adjustment on Thermostat

Dip Tube

Upper Element

Lower Element

Courtesy A.O. Smith, Corp.

Fig. 11-2. Cutaway view of A.O. Smith electric water heater.

Troubleshooting Electric Water Heaters

Nature of Trouble	Possible Cause	Service Procedure
No hot water	1. Fuses out or blown, or circuit breakers switched off 2. Element burned out 3. ECO has opened 4. Timer has turned electricity off	1. Replace fuses, turn circuit breakers on. 2. Call service technician. 3. Press reset button on thermostat. 4. If special meter for water heater is controlled by timer, recovery is limited to certain hours. If owner's timer has turned electricity off, wait for timer to reset to ON or turn timer manually to ON.
Insufficient hot water	1. Heater undersized	1. Reduce rate of water use. If heater is operated by manually adjusted timer, set timer for longer operating periods.
Water is too hot or too cold	1. Thermostat not set for desired water temperature	1. Adjust thermostat located under cover plate on jacket of heater.

Caution: For your safety, *do not* attempt to test or replace thermostat or element; refer repairs to qualified service personnel. 240 volts (±10%) are present in electrical panels, timers, thermostats, and elements. Use extreme caution when replacing fuses, switching circuit breakers, adjusting timers, or resetting thermostats. *When resetting thermostats, do not remove protective fiber cover* over thermostat.

Note: If pressing the reset button on the thermostat does not result in restoring hot-water supply within a few hours, indications are that the element is burned out. Call service technician.

larly from an electric water heater. If the sediment is allowed to build up and cover the heater element, it will burn out.

Draining procedure—Turn off the valve on the cold-water supply to the heater, open the drain valve and allow water to flow until it is clear. Opening a hot-water faucet prevents a vacuum from forming during the draining process. If the drain valve becomes clogged with sediment, open the cold-water supply valve momentarily; the pressure applied should clear the valve.

Plastic drain valves that are installed on many water heaters often leak after use and cannot be shut off. This valve can be removed and a ¾" I.P.S. (iron pipe size) gate valve with a ¾" × 3" long galvanized nipple in one end and a ¾" hose adapter on the other installed in its place. A rod can be inserted through the open gate valve to clear any obstruction blocking the drain opening and the hose connection will allow a garden hose to be used to conduct the water away from the heater. An anode, shown in Fig. 11-1, is used on water heaters to prolong the life of a glass-lined tank. This is called a cathodic protection process, and the rod is slowly consumed during this process. The anode rod should be removed and inspected periodically and replaced when more than 6" of core wire is exposed at either end of the rod.

When water containing a high percentage of sulfate and/or other mineral content is heated, it can produce a hydrogen sulfide or "rotten egg" odor. Removing the magnesium anode may help this problem, but removal of the anode may also void the warranty on the tank. Water treatment should minimize the odor problem. *Caution:* Hydrogen gas can be produced in a hot-water system that has not been used for a period of time, two weeks or more. *Hydrogen gas is extremely flammable.* To reduce the risk of injury under these conditions, it is recommended that the hot-water faucet on the kitchen sink be opened for several minutes before using any

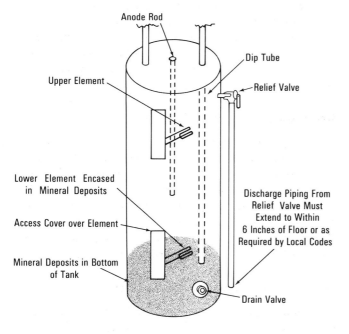

Fig. 11-3. Mineral deposits cause burned-out elements.

electrical appliance connected to the hot-water system. If hydrogen is present, it may sound like air escaping when the faucet is opened. There should be no smoking or open flame near the faucet while it is open.

Inspecting the Heater and Vent Piping

The internal flueway of a gas water heater and the vent piping and draft from a heater should be inspected annually to be certain that they are clean, open, not rusted out, and a good draft is present. The valve on the gas supply to the heater must be in OFF position during this inspection. The draft diverter should be removed and the flue baffle lifted out. After inspecting the internal flueway and removing any soot, scale, etc., reinstall the flue baffle, making certain the baffle is hung securely by its hanger. Inspect the burner chamber and remove soot, lint, or other debris that can interfere with proper combustion. If soot is found, it is an indication of improper combustion. The heater should be turned OFF and left OFF until qualified technicians can investigate and remedy the problem. The vent piping should be carefully checked for rust spots, holes, or weak spots that would allow combustion products (carbon monoxide, etc.) to escape into the house. Carbon monoxide is odorless and deadly.

After careful checking, if no problems are found, all parts removed during the inspection must be replaced, and the heater can be placed in service.

One more test should be made when the main burner is on. Check for a good draft in the vent piping by holding a lighted match near the diverter lip, as shown in Fig. 11-1. If the flame and smoke from the match are drawn up into the diverter, this indicates a good draft. If the flame or smoke are pushed out away from the diverter, the draft is poor, allowing combustion products to escape into the area. The heater should be turned off immediately, and then a qualified person should be called in order to correct the problem.

Combustion and Ventilation Air

Proper operation of a gas water heater requires air for combustion and ventilation. If the water heater is installed in an unconfined space within a building of conventional frame, masonry, or metal construction, infiltration air is normally adequate for proper combustion and ventilation. When a gas heater and a gas furnace are installed in the same room, if the room is tightly closed with little air infiltration, a fresh air inlet may be required.

CHAPTER 12

How to Install or Reset a Toilet

Repair or remodeling projects may require taking up and resetting a toilet or installing a new one. When taking up a toilet, the tank should be removed from the bowl first. Before this can be done the valve (Fig. 12-2) must be shut off and the toilet flushed to remove most of the water in the tank. The balance can be sponged out. If the shut-off valve leaks, water will enter the tank through the fill valve. If this happens, the shut-off valve must be replaced. When taking up a toilet, the tank should be removed from the bowl. Fig. 12-1 shows the bolts extending through the tank with nuts and washers under the top edge of the bowl. These nuts must be removed and coupling nut "A" (Fig. 12-2) must be disconnected from the fill valve shank. The tank can then be lifted up and off of the bowl. Fig. 12-1 shows how the bowl is secured to the closet flange and the floor by closet bolts inserted under the closet flange. When the nuts on the closet bolts are removed, the bowl can be lifted off the flange. When a toilet that has been in use for several years is taken up, it is usually necessary to use a new gasket between the tank and the bowl. The local plumbing shop should be able to match the old gasket.

When resetting a toilet or installing a new one, new closet bolts should be used. Closet bolts are made in ¼″ and ⁵⁄₁₆″ sizes; ⁵⁄₁₆″ being the heaviest, is much the best.

SETTING AND CONNECTING THE TOILET

Insert the bolts under the slots or through the holes in the closet flange. Set the closet bowl on end and slide the wax ring over the horn on the closet bowl, as shown in Fig. 12-3. Holding the bowl by the rim, pick it up and set it carefully on the closet flange. When the bowl is centered over the closet flange, the closet bolts will protrude through the holes in the bottom of the bowl; push down on the bowl to "set" the wax ring. Drop the chrome-plated washers on the bolts and start the open nuts on the bolts. Screw the nuts down *hand tight*. If you're setting a new toilet, follow the assembly instructions furnished with it. If you're resetting an old one, insert the tank bolts with a rubber washer under the heads, through the holes in the tank. If a new gasket is used between the tank and the bowl, press it over the flush valve end on the bottom of the tank. Holding the tank, sit down on the bowl, facing the wall in back of the toilet. Set the tank down on the bowl and, while pushing down on the bolt head with one hand, slide the rubber washer, flat steel washer, and lock washer over the bolts and start the nuts on the bolts with the other hand. A small adjustable wrench can be used to tighten the nuts. A long screwdriver can be used to hold the bolt heads while tightening the nuts

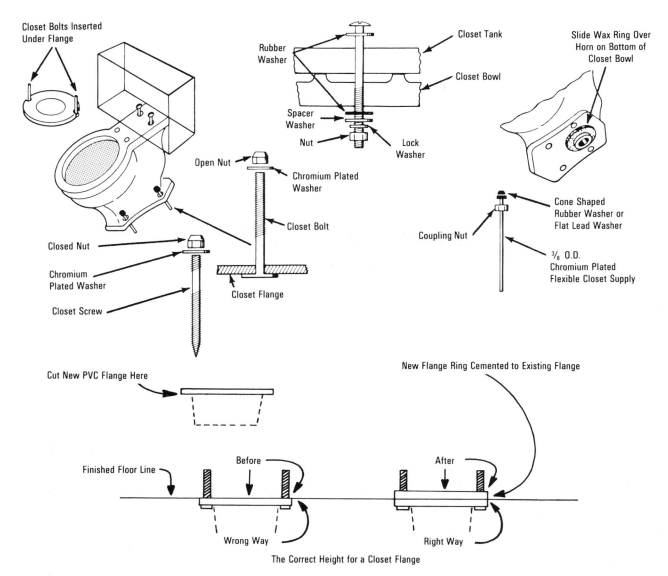

Fig. 12-1. **Details for taking up and resetting a toilet.**

at the bottom. Tighten the nuts evenly, going back and forth, as the nuts get tight turn them only a quarter turn at a time, until the tank is solidly mounted on the bowl. As the bolts are being tightened, the tank must be level. Some closet bowls have raised places on the top where the tank mounts. When the tank touches these raised places, *stop tightening the nuts.* Further tightening of the nuts will break either the tank or the bowl.

The closet bolt nuts at the floor can now be tightened again, hand-tight first, then using a wrench, tightened no more than one turn each. Extreme care must be taken when tightening these nuts. The bowl is vitreous china, porcelain, and can be easily broken. If the above instructions have been followed, the tank and bowl should be securely mounted to the floor and the closet bolt nuts can be retightened, if necessary, a

day or two later. After setting the toilet and tightening the bolts, grasp the bowl by the rim and try to move it from side to side. If it is solid, the bolts are tight enough. If china caps are to be placed over the nuts, the threads above the nuts must be sawed off with a fine-toothed hacksaw. If sawing the bolts off loosens the nuts, retighten them.

CONNECTING THE WATER SUPPLY

A flexible closet supply tube is used to make the connection to the fill valve in the tank. The flexible supply is made of soft copper tubing and can be bent or shaped to fit between the compression stop valve and the fill valve shank. If a short, sharp offset is

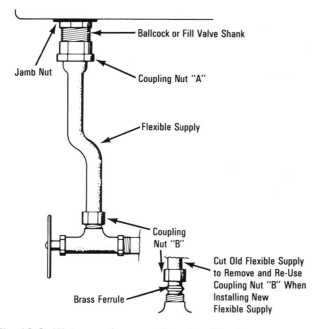

Ballcock or Fill Valve Shank

Jamb Nut

Coupling Nut "A"

Flexible Supply

Coupling Nut "B"

Cut Old Flexible Supply to Remove and Re-Use Coupling Nut "B" When Installing New Flexible Supply

Brass Ferrule

Fig. 12-2. Water supply connections to a fill valve.

needed, the supply tube can be inserted into a ⅜″ O.D. bending spring and the tube can be shaped without kinking it. A bending spring and flexible supply tube are shown in Fig. 12-3. Flexible closet supplies are available wherever plumbing supplies are sold.

CORRECTING LEAKS AT THE FLOOR LINE

The prime cause of leaks at the floor line around a toilet bowl is a closet flange that is set too low. The flange may be too low because it was not set correctly when the house was built or because additional floor covering, tile, etc., has been added. For a watertight seal, the *top* of the flange must be at the right height. This means that the *bottom edge* of the flange must set *on* the finished floor, as shown in Fig. 12-1. Two wax rings, instead of one, are sometimes used in an attempt to make up for a flange that is set too low. This usually

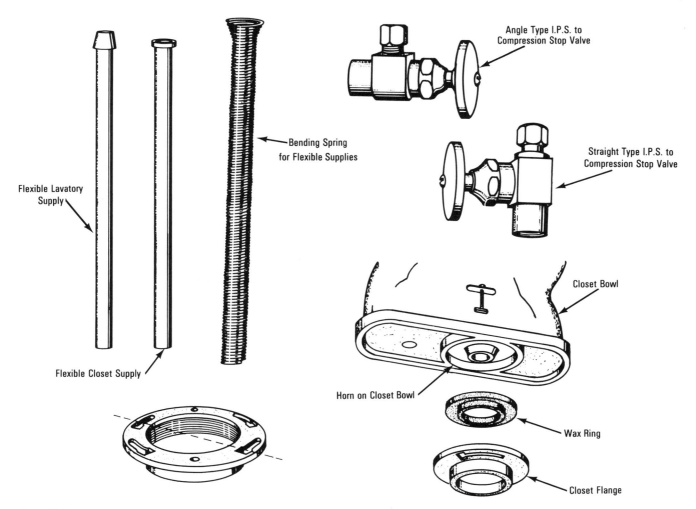

Bending Spring for Flexible Supplies

Flexible Lavatory Supply

Flexible Closet Supply

Angle Type I.P.S. to Compression Stop Valve

Straight Type I.P.S. to Compression Stop Valve

Closet Bowl

Horn on Closet Bowl

Wax Ring

Closet Flange

Fig. 12-3. Parts used when setting a toilet.

results in wax being continuously squeezed out from under the bowl, creating an unsightly condition and resulting eventually in leaks. If a flange is too low and is made of cast iron, a plumber should be called in to install a new flange at the correct height. If the flange is made of PVC, the top edge of another flange can be sawed off, as shown in Fig. 12-1, and cemented, using PVC cement, on top of the original flange. The slots and holes in the cemented flange must line up with the ones in the original flange.

CHAPTER 13

Repairing and Replacing Sink and Lavatory Traps

KITCHEN SINK TRAPS

One of the most common repair jobs in the home is the repair or replacement of kitchen sink and lavatory traps. Not too many years ago replacement of a kitchen sink trap was a difficult job, requiring the use of large pipe wrenches, usually a large hammer to break a cast-iron fitting which wouldn't turn, and a good supply of choice expletives to mutter under one's breath when all else failed. Thanks to new materials and techniques, those days are gone, at least in fairly new houses. And in older houses, the cast-iron drainage fittings, cast-iron traps, etc., can be replaced with PVC traps, piping, and drainage fittings. Replacing a leaking trap can be as simple as loosening and removing two slip nuts, taking out the old trap, inserting the new trap, and tightening up two new slip nuts.

Fig. 13-1(A) shows a trap connected to the drainage piping through a PVC-DWV compression to female slip adapter. Different piping arrangements might have required the use of (B), a compression to F.I.P. (female iron pipe) adapter, to (C), a compression to male slip adapter, or (D), a compression to M.I.P. (male iron pipe) adapter. Adapters to fit galvanized piping, copper tubing, or PVC drainage piping are available at stores selling plumbing material.

Fig. 13-1(E) shows a P trap with a cast brass ell (elbow). This type of trap is made with threads on the inside of the cast ell, to fit a male pipe thread. To replace this type trap, the adapter (B) and the tube P trap would be used.

Quite often a cracked or leaking J bend of the types shown in (F) and (G) can be replaced without changing any of the drain piping. Replacement J bends are readily available wherever plumbing supplies are sold.

LAVATORY TRAPS

Replacement of a lavatory trap can often be a problem. There are several methods of connecting traps to the drainage piping.

1. In a fairly new house the trap will probably be connected by one of the types of trap adapters shown in Fig. 13-1. Replacing a trap connected in this way should be no problem. Loosen and remove the slip nuts on the inlet and outlet of the trap, cut the trap to fit if necessary, set the trap into place, and tighten the slip nuts.

2. In a home with DWV copper drainage piping, the trap arm may have been inserted into the end of the tubing and soldered to the tubing at this point. To remove the old trap, heat the soldered joint with a propane torch (if the joint is near the

117

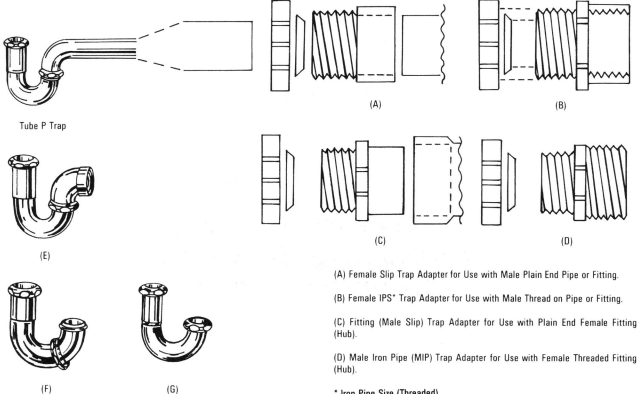

Tube P Trap

(A)

(B)

(C)

(D)

(E)

(F)   (G)

(A) Female Slip Trap Adapter for Use with Male Plain End Pipe or Fitting.

(B) Female IPS* Trap Adapter for Use with Male Thread on Pipe or Fitting.

(C) Fitting (Male Slip) Trap Adapter for Use with Plain End Female Fitting (Hub).

(D) Male Iron Pipe (MIP) Trap Adapter for Use with Female Threaded Fitting (Hub).

* Iron Pipe Size (Threaded)

Fig. 13-1. Connecting a trap with a trap adapter.

wall, be careful not to start a fire). When the solder melts, use a pair of pliers to pull the old trap out. Any burrs or sharp edges should be filed from the end of the copper drainage tubing and the outside of the tubing should be cleaned with sandpaper or sand cloth about 1″ back from the end. Chapter 6, **How to Work with Copper Tubing**, explains how to prepare a joint for soldering. Measure the I.D. (inside diameter) of the tubing; a trap adapter from the I.D. tubing size to 1¼″ O.D. (outside diameter—the trap size) will be needed. The adapter must be soldered to the cleaned end of the drainage piping, and the trap can be cut to fit and replaced. Remove the slip nut and plastic washer (if used) from the adapter while soldering it to the tubing.

3. In some areas, until fairly recently, lead pipe was used for drainage pipe and traps were soldered into the lead pipe at the wall line. Replacing a trap that was installed this way is a job for a highly skilled craftsman. If a trap that needs to be replaced is soldered to a lead drain pipe, now might be the time to replace not only the trap but also the lead pipe.

4. You may find that a trap that is leaking and must be replaced is connected to the drain piping at the wall by a brass solder bushing. The trap is soldered to the brass bushing, which has a male pipe thread, and the bushing and trap arm is then screwed into the drainage fitting. A solder bushing that has been in a drainage fitting for several years is almost impossible to remove except by cutting it out. Fig. 13-2(A) shows a trap arm soldered into a solder bushing, which in turn is screwed into a drainage fitting. A hammer and a small chisel or old screwdriver can be used to cut the trap arm out of the solder bushing, leaving the bushing as shown in Fig. 13-2(B). A hacksaw blade can be used to make the two cuts shown in (B). Saw through the brass bushing only; do not saw into the fitting threads. Brass is brittle; the small chisel can be used to knock out the piece between the saw cuts. With this piece out, the rest of the bushing can be easily removed.

With the old trap and solder bushing removed, the trap can be replaced. With a female pipe thread (fitting) at the wall, a male adapter (D) would be needed and a PVC adapter could be used at about one-fourth the cost of a brass adapter. A chrome-plated brass P trap could be used if the trap is exposed. If the trap is concealed in a cabinet, a PVC trap could be used.

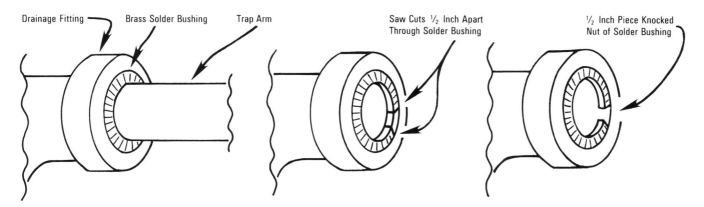

Fig. 13-2. Removing a solder bushing.

When measuring pipe sizes, remember that plumbing pipe and fittings are measured by I.D. (inside diameter). Thus, a 1½″ × 1¼″ compression adapter would have a male thread measuring approximately 1⅞″ O.D. (outside diameter).

Before starting the adapter in the fitting, wrap one layer of Teflon tape dope on the threads, beginning at the end of the thread and wrapping in the direction of the threads.

CHAPTER 14

Installing a New Lavatory and Vanity Cabinet

If you would like to improve the appearance of your bathroom and at the same time gain much-needed storage space, replace the old lavatory with a new vanity cabinet and a new lavatory. It's a weekend project that can add far more than its cost to the value of your home. Vanity cabinets and tops are available in a wide variety of styles, colors, and sizes to fit every bathroom. Cabinet tops can be of laminated plastic (Formica, etc.) into which a lavatory can be set; "cultured marble" tops with precast lavatories are also available. The first step in this project is to measure the available space and select the vanity and lavatory. A cabinet that will center on the space occupied by the old lavatory will be the easiest to install. The cabinet should be open, with no drawers, shelves, or spacers in the space needed for water and drain connections.

The old lavatory must be disconnected and removed. The hot and cold valves under the lavatory must be shut off; if they will not shut off completely, they should be replaced. This will probably require shutting off all the water in the house while the valves are being replaced. The water supply connections to the faucet will be one of the three types shown in Fig. 14-2(A), (C), or (D). If the connections are as shown in (A), a basin wrench, Fig. 1-16, will probably be

needed to loose and remove the coupling nut. Two adjustable wrenches, one as a back-up, can be used to loosen the connections shown in (C) or (D). Next, disconnect the trap, and the lavatory can be lifted off its brackets and removed. The remaining drain piping back to the wall should be removed; the instructions in Chapter 13 will be helpful. Fig. 14-4 shows two suggested methods of connecting the trap from the new lavatory to the F.I.P. (female iron pipe) fitting in the wall. The change from iron pipe to PVC-DWV pipe is made by using a PVC-DWV M.I.P. (male iron pipe) adapter. Before setting the new vanity in place, the drain opening at the wall should be extended to a point where it will be accessible inside the cabinet. If the existing drainage piping is copper tubing, the trap may be connected to a copper sweat adapter (Fig. 13-1). This adapter can be removed and a copper to PVC-DWV adapter (Fig. 8-3) can be used. If no adapter is used, the trap will be soldered to the copper tube. Either soldered joint can be heated and the adapter or trap removed. Use caution here to avoid starting a fire in the wall. Any burrs or ridges on the end of the copper tube should be filed off before the copper to PVC adapter is pushed on. Two O-rings inside the adapter make a watertight joint.

Fig. 14-1. For appearance and convenience add a new vanity.

The back of the cabinet must be cut out for the water and drainage piping, and the cabinet can be set in place and secured to the wall. The exact fastening method will depend on the wall construction. Ordinarily a wood screw or toggle bolt in each corner of the back, top and bottom, will anchor the cabinet securely. Before the screws or bolts are tightened, the cabinet should be leveled, using shims under the base. A strip of molding can be used to cover any gaps at the floor line caused by leveling.

SETTING THE LAVATORY

Vitreous china counter-top self-rimming lavatories are the most popular type. They are usually centered in the length of the counter top and with the front edge approximately 1½" from the front edge of the top. The exact placement will depend on how the top is made. A template should be packed in the lavatory shipping box with instructions for its use. *Follow these instructions carefully*; the counter top can be ruined if the hole is cut in the wrong place. An electric saber saw can be used to cut the hole. After the hole is cut the lavatory should be set in place to make certain it fits properly. After this has been determined, the lavatory can be set permanently. A tube of setting compound should be packed with the lavatory; this compound should be applied around the bottom edge of the rim of the lavatory as shown in Fig. 14-3. If the lavatory is furnished without setting compound, *Polyseamseal All-Purpose Caulking* is a good setting agent, and any excess can be cleaned up easily using a damp cloth. When the lavatory has been set in place, it should be checked carefully to make sure it is positioned correctly. After the setting compound has set up, the lavatory can not be moved. The setting compound should be allowed to set up at least two or three hours before connecting the water and waste piping to the lavatory.

SETTING AND CONNECTING THE LAVATORY FAUCET AND POP-UP DRAIN

The carton containing the new faucet and pop-up drain will also contain assembly instructions. These instructions may not specify the application of a ring of soft plumber's putty under the drain spud of the lavatory. If they do not, you should apply the putty as shown in Fig. 14-3 to prevent water leakage around the spud. PVC traps and piping are replacing chrome-plated brass tubing for drainage connections where piping is concealed in cabinets. PVC piping, 1½", is connected to 1¼" chrome-plated tubing used for traps, etc., by using a 1½" × 1¼" plastic slip nut washer under the slip nut at the inlet of the trap. Fig. 14-4 shows the various fittings that will be needed for either an offset or a straight connection from drain piping in the wall to the lavatory trap. A chrome-plated tubing trap can be used if desired by using a 1½" × 1¼" plastic slip nut washer under the slip nut at the trap adapter. The cutting and cementing of PVC piping and fittings is explained in Chapter 8.

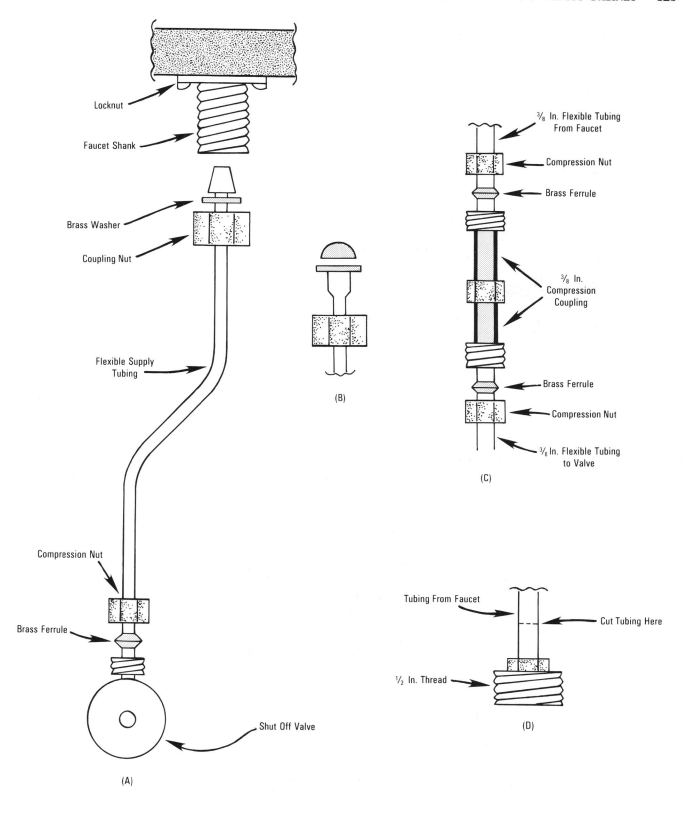

Locknut

Faucet Shank

Brass Washer

Coupling Nut

Flexible Supply
Tubing

Compression Nut

Brass Ferrule

Shut Off Valve

(A)

(B)

⅜ In. Flexible Tubing
From Faucet

Compression Nut

Brass Ferrule

⅜ In.
Compression
Coupling

Brass Ferrule

Compression Nut

⅜ In. Flexible Tubing
to Valve

(C)

Tubing From Faucet

Cut Tubing Here

½ In. Thread

(D)

Fig. 14-2. How to use ⅜″ flexible tubing for connections.

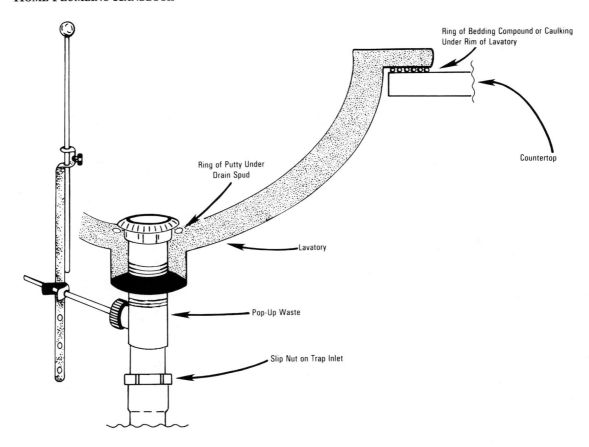

Fig. 14-3. Setting lavatory and installing pop-up waste fitting.

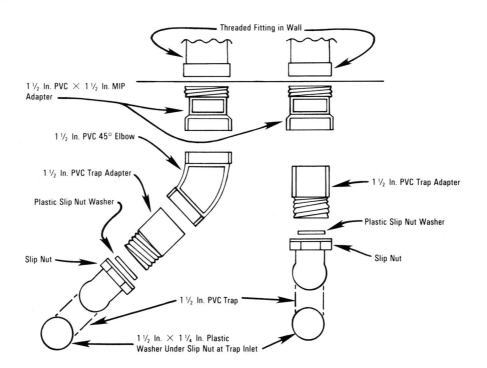

Fig. 14-4. Offset and straight drainage connections.

CHAPTER 15

Caulking Around Plumbing Fixtures and Tile

Water leaks through wall surfaces can seriously damage the dry-wall or plastered wall beneath the wall surface. Counter tops are almost universally made of particle board. Particle board is strong and tough, but it is not water resistant; it literally melts and crumbles away if water contacts it. A good caulking carefully applied will seal out water and prevent damage to walls and counter tops. Water leakage through walls is most prevalent around bathtub edges and bathtub faucets and spouts. Water leakage also occurs at wall corners, along the front (apron) of the tub, and around recessed soap and grab fixtures. I have tried many types of caulking materials, and the best by far is *Poly-seamseal All-Purpose Caulking*. It is white and stays white and is easily applied using a caulking gun, as shown in Fig. 15-1. It is available at building supply stores and seems to keep indefinitely, even after opening. Directions for its use are printed on the tube, and when applied as directed it will prevent water leakage at wall joints, cracks, fixtures. It is also excellent for use as a bedding compound, used under the rim, when setting self-rimming sinks and lavatories. Best of all, *Polyseamseal* is a water-clean-up type of caulking if the clean-up is done before the caulking "sets" up.

Caulking used as a bedding agent is shown in Fig. 14-3.

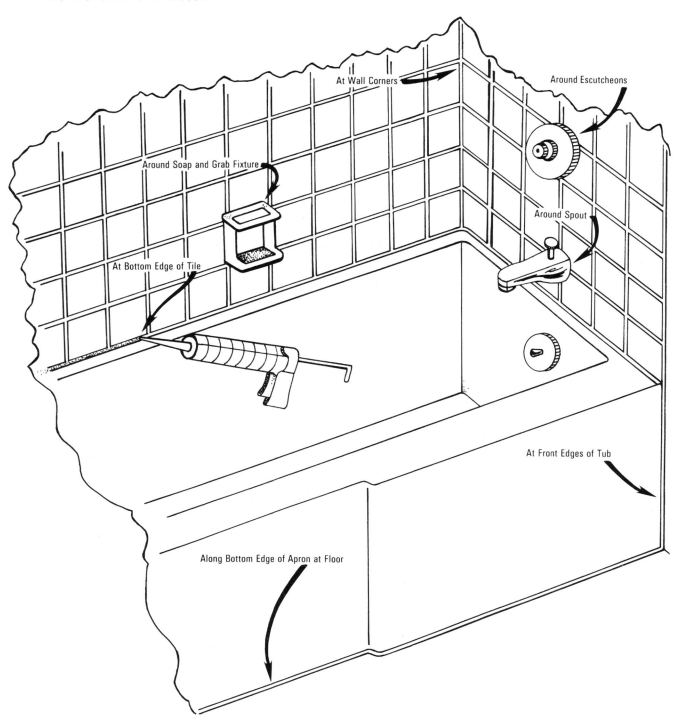

Fig. 15-1. Where to use caulking to prevent leaks.

CHAPTER 16

Installing a
Homemade Humidifier

An effective humidifier is a necessity in today's homes. The coldest weather in winter is produced by high-pressure areas sweeping in from Arctic regions. This Arctic air is very cold and extremely dry. In homes heated by hot-air furnaces, the cold dry air is heated and circulated through the home. Furniture with glued joints comes apart. Static electricity is formed when walking across certain types of carpet and a shock is felt when metal is touched. Doctors often advise the installation of humidifiers to alleviate the ill effects of dry air on the human body.

A great many homes today are built on a concrete slab and the plumbing pipes and the heating air supply ducts are located under the concrete slab. The hot-air furnace used with this type of construction is called a counter-flow furnace. When a counter-flow furnace is used, the hot air supply ducts are located under the floor and the return or cold air ducts are located above the ceiling. When the home is of this type of construction, a homemade trouble-free humidifier can be installed in the hot air supply ducts for a fraction of the cost of conventional humidifiers. The humidifier can be manually or electrically operated to keep the humidity within the desired range. In order to know how to make the necessary electrical connections, it is first necessary to understand the sequence of actions in the operation of the furnace.

For the purpose of this explanation, we will assume that the furnace is equipped with a gas burner. The actual operation is essentially the same if the furnace is equipped with an oil-fired burner. When the room thermostat senses a drop in temperature, below the set level, a switch in the thermostat closes. (When we say a switch in the thermostat closes, we mean that an electrical circuit is completed, and current flows in the circuit.) An open switch interrupts the flow of current in that circuit. When the thermostat switch is closed, electrical current flows in the circuit to the gas valve in the furnace, the gas valve is energized, and the gas valve is turned on. The main burner then lights and heat builds up in the heat exchanger. Many furnaces have a combination fan-limit switch to control the operation of the fan and the bonnet/heat exchanger temperature of the furnace. When the temperature of the air in the heat exchanger has reached the set level, the fan switch closes and the fan is turned on. The cooler air is pulled back from the rooms through the cold air return ducts, and pushes the heated air back into the rooms through the hot air supply ducts.

When the temperature in the heat exchanger reaches the setting of the limit switch, the limit switch is opened and the electrical current to the gas valve is cut off, thus shutting off the gas burner. The fan continues to operate, circulating the hot air back to the rooms in

the house until the desired setting of the thermostat is reached. Most thermostats have a heat-anticipating feature built into them. This feature regulates the point at which the fan will be turned off. The operation of the humidifier described in Fig. 16-1 requires only two electrical connections. The two wires from the sole-

noid valve must be connected to the two wires going to the fan motor. Thus, when the fan motor is energized, the solenoid valve is also energized; the solenoid valve is opened and water will spray out of the tubing into the hot air ducts under the furnace. The hot dry air from the furnace will pick up the moisture from the

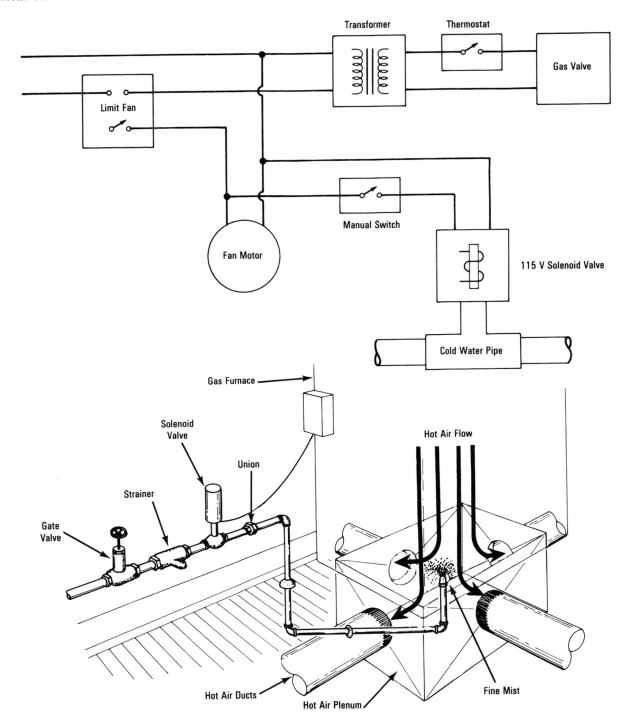

Fig. 16-1. Construction of a homemade humidifier and the electrical connections.

water spray and raise the humidity of the air to the desired level. When the humidity is at the desired level, the manually operated switch shown in the illustration can be turned off. This interrupts the circuit to the solenoid valve, shutting it off but leaving the fan in operation. When the humidity falls below the desired level, the switch can be turned back on, putting the humidifier back into action.

If a more sophisticated system is desired, a humidistat can be installed to control the operation automatically. Wall- or duct-type humidistats are available for this purpose. The humidistat will sense the moisture content of the air, and when it falls below the desired level, a switch in the humidistat will close (completing an electrical circuit to the solenoid valve), turning on the solenoid valve. When the humidity reaches the desired level, the switch in the humidistat opens, the circuit to the solenoid valve is also opened, and the solenoid valve is shut off.

The fan voltage is normally 120 volts, so the solenoid valve must also be rated for 120-volt operation. If a line voltage (120 volts) humidistat is used, the proper size of wire must be installed from the solenoid valve to the humidistat. Local electrical codes must be followed when installing this wiring. The author recommends that a qualified electrician install necessary wiring and also make the required connections. For either of these installations, the solenoid valve must be a normally "closed" type; that is, when there is no electrical current to the coil on the valve, the valve is shut off. When the coil is energized, the valve opens. To prevent any scale or other foreign matter from lodging in the solenoid valve, a strainer should be installed in the pipe, as shown in the illustration. Solenoid valves

and strainers are available at refrigeration supply stores.

Installation of the pipe from the solenoid valve to the plenum chamber under the furnace will of necessity depend on the manner in which the furnace is installed. It is here that the ingenuity of the homeowner will be used. The pipe from the solenoid valve should extend into one of the heating ducts leading from the plenum boxes. The pipe must be removable; the tip end will occasionally develop a formation of lime or scale and have to be cleaned. A larger pipe could serve as a sleeve through which the humidifier pipe would enter the plenum. It would then be a simple matter to loosen the union in the humidifier pipe and remove the pipe for inspection. The pipe is assembled as shown in Fig. 16-1. The end of the soft copper tubing is pinched shut with pliers. Pinching the tubing shut will not keep water from coming out under pressure; it will force the water out as a fine spray or mist. the hot, dry air from the furnace will absorb the moisture from the spray and raise the humidity level in the home.

Galvanized steel piping with galvanized malleable iron fittings could of course be used instead of copper tubing, as shown in the illustration. The pip inserted through the sleeve and into a duct opening should be $\frac{1}{8}''$ I.D. ($\frac{1}{4}''$ O.D.) soft copper tubing. If galvanized pipe and fittings are used up to the gate valve, strainer and solenoid valve use a $\frac{1}{2}''$ male iron pipe to $\frac{1}{4}''$ O.D. ferrule-type compression adapter, screwed into the solenoid valve to adapt from iron to copper pipe. If this fitting is used, since it is in effect a union, no other union is necessary at this point.

CHAPTER 17

Water Conditioning

There are many areas where the installation of a water conditioner is very desirable. While all water contains (among other elements) calcium, in certain areas the water has a very high calcium content. This condition, called "hard" water, is easily overcome by the use of a water softener. Iron is often present in the water obtained from wells; in some cases the iron content is extremely high. When the water is pumped into a tank and exposed to air, the iron content is changed to an iron oxide. The iron oxide will form thick deposits in pressure tanks, piping, water closet tanks, etc., and eventually clog the piping. Where this condition is encountered, an iron filter should be installed. The water treatment sequence would be (1) through the iron filter, (2) through the water softener. It is not necessary to soften all the water used in the home. The water used for drinking, cooking, water closets, and sprinkling purposes need not be softened. Common practice is to soften only the hot water.

Note that a bypass connection is shown in Fig. 17-1. This is desirable even if the water softener has a built-in bypass connection. In the event the water softener needs repairing or has to be replaced, hot water can still be obtained if a bypass connection is correctly installed in the piping system.

Water is softened by circulating it through Zeolite. Zeolite is a manufactured resin, made in very small beads that resemble an orange-colored sand. Zeolite catches the magnesium and calcium compounds that make the water hard. Periodic backwashing with salt water restores the Zeolite to its original condition. Well water with a high hydrogen sulphide (sulfur) content, causing a "rotten-egg" smell, can be treated to remove the sulfur and the odor. Several methods are used to treat this type of condition, depending on the concentration of hydrogen sulphide and iron in the water. Neutralizing filters can be used to correct acid water conditions. Companies that handle water softener equipment also handle other water treatment supplies.

Water that has been softened often has a high salt content. This occurs because the mineral is backwashed with brine and the brine is not completely removed by the backwash process. Many doctors have advised their patients not to drink softened water. In the piping arrangement in Fig. 17-1(A), only the hot water would be softened. In Fig. 17-1(B), hard (unsoftened) water would be supplied to the cold-water opening at the kitchen sink, to toilets, and to sillcocks.

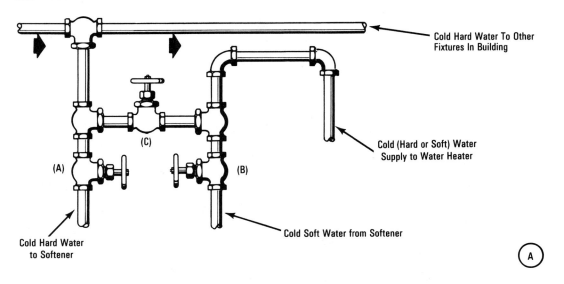

Cold Hard Water To Other
Fixtures In Building

Cold (Hard or Soft) Water
Supply to Water Heater

(C)

(A)   (B)

Cold Soft Water from Softener

Cold Hard Water
to Softener

(A)

In Normal Operation: Valve A is Open
Valve B is Open
Valve C is Closed

Cold Hard Water Enters Softener Through Valve A
Leaves Through Valve B

If Softener Needs Repair, Valve A is Closed
Valve B is Closed
Valve C is Open

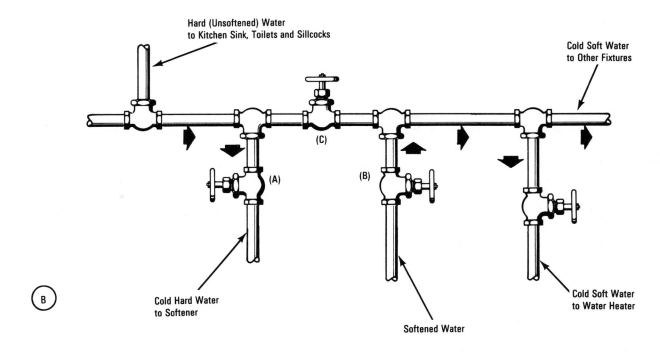

Hard (Unsoftened) Water
to Kitchen Sink, Toilets and Sillcocks

Cold Soft Water
to Other Fixtures

(C)

(A)   (B)

Cold Soft Water
to Water Heater

Cold Hard Water
to Softener

Softened Water

(B)

If it is Desirable to Use Hard (Unsoftened) Water to Supply the
Water Closets and Sillcocks and Use Softened Water for the
Other Fixtures, the Piping Connections Could Be as Shown in
Fig. A.

The Bypass Connection Operation Would Be: Valve A, Normally Open
Valve B, Normally Open
Valve C, Normally Closed

In the Event of Repair or Replacement of the Water Softener,
Valve A Would Be Closed
Valve B, Would Be Closed
Valve C Would Be Open

And a Supply of Hot Water (Unsoftened) Could Be Maintained
Until the Water Softener is Repaired or Replaced.

Fig. 17-1. Various plumbing connections for a water softener.

CHAPTER 18

Working with Cast-Iron Soil Pipe

Cast-iron soil pipe is the most durable material available for drainage waste and vent piping. Special tools are needed when working with soil pipe. Cold chisels or soil pipe cutters may be used to cut cast-iron soil pipe to the desired length. A yarning iron, a packing iron, and inside and outside caulking irons are needed to make the joint between two pieces of soil pipe. A lead melting furnace is used to heat the lead pot. When the lead is hot, a ladle is used to pour lead into the joint. When horizontal joints are made, a joint runner is used to hold the molten lead in the joint until the lead has solidified. Molten lead is dangerous to work with, but some simple precautions will minimize the danger.

Never add lead that is wet or even slightly damp to a pot of melted lead. *Never* put a wet or even slightly damp ladle into a pot of melted lead. Hold the ladle above the pot of the melted lead until it is warm and dry. When lead has to be added to the pot, lay a cake of lead in the ladle and hold the ladle and cake of lead over the pot until all moisture has evaporated from the ladle and the lead. If a damp or wet ladle or cake of lead is inserted into the molten lead in the pot, some of the moisture will be carried under the surface of the molten lead. The moisture will be instantly converted into steam. The steam then literally explodes, throwing hundreds or thousands of particles of molten lead in all directions. Severe burns and loss of eyesight are just

two of the serious consequences that could occur in such an accident.

Never pour melted lead into a wet joint or a joint in which the oakum is wet or damp. Again, water under the hot lead will convert into steam and an explosion could result. Use only oiled or tarred oakum in soil pipe joints. Oiled or tarred oakum resists moisture.

Cast-iron soil pipe is made in standard lengths. S.H. (single hub) is made in 5' and 10' lengths. Measurements for cut lengths should be end-to-center measurements. The spigot end is inserted into the hub, the end measurement is taken from the point where the spigot end meets the bottom of the hub.

In Fig. 18-1, a piece of soil pipe must be 42″ from end (of pipe) to center of ¼ bend. The 4″ ¼ bend measures 8″ from end to center. Therefore, 42″ (total end-to-center length) minus 8″ (end to center of the ¼ bend) equals 34″. The piece of pipe must be cut 34″ long. An end-to-center measurement of a wye (Y) or a bend (⅛ - ⅙ - ⅕ - 1⁄16 ″) is made as shown in Fig. 18-1.

Cast-iron soil pipe is used extensively for drainage, waste, and vent pipe. Cast iron has the characteristic of rusting on the surface but protecting itself against the continued rusting which would in time destroy the metal. The initial coat of rust adheres strongly to the metal and prevents further rusting action. Cast-iron pipe and fittings can be joined in several different ways, but the conventional way is to use lead joints.

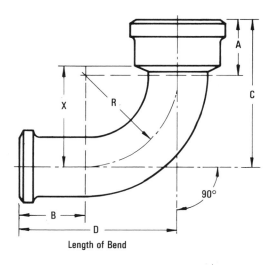

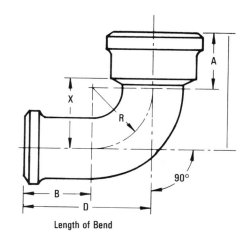

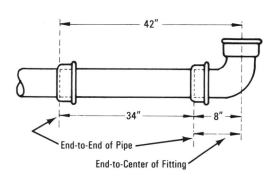

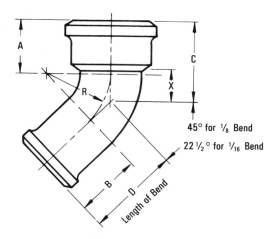

Fig. 18-1. How to measure bends.

When two pieces are to be joined, the spigot end of one piece is set into the hub of the other piece.

Oakum is generally sold in individual boxes of 5 lbs. each; the strands of oakum are usually approximately 2′ in length. Several strands are twisted together; the oakum should be untwisted and used one strand at a time. Place one end of the strand at the top of the hub and push the oakum down into the hub with the yarning iron. Follow the strand of oakum around the hub, ramming the oakum into place in the hub, using the yarning iron. Oakum should be rammed into the hub, leaving a 1″ space between the top of the oakum and the top of the hub. The oakum should be packed tight in the hub, using a hammer to drive the packing iron for final packing. The lead is then poured into the joint; the joint should be filled to the top of the hub in one pour. The molten lead will solidify in a very few seconds. When the lead has solidified, use a hammer and inside caulking iron and go around the inside of the joint. When the inside of the joint has been caulked, use the outside caulking iron and go all around the outside edge of the joint.

There are two important points to remember when caulking a joint: (1) Use a lightweight hammer, not over 12 oz., and do not hit the caulking iron too hard. Lead is soft, and when too much force is used on the caulking iron, the lead will be expanded against the outside edge of the joint. The expansion of the lead can crack the cast-iron hub. (2) The oakum, not the lead, is the key to a watertight joint; the old joke among plumbers, "Do a good job, use lots of oakum" (lead was much more expensive than oakum) was not really a joke after all. If the oakum is tightly packed in the joint, the joint will be waterproof, and the lead only serves to hold the oakum in place. See Fig. 18-3.

See Fig. 18-4:

(A) Yarning iron—used to insert oakum into the space between the spigot end and the hub.

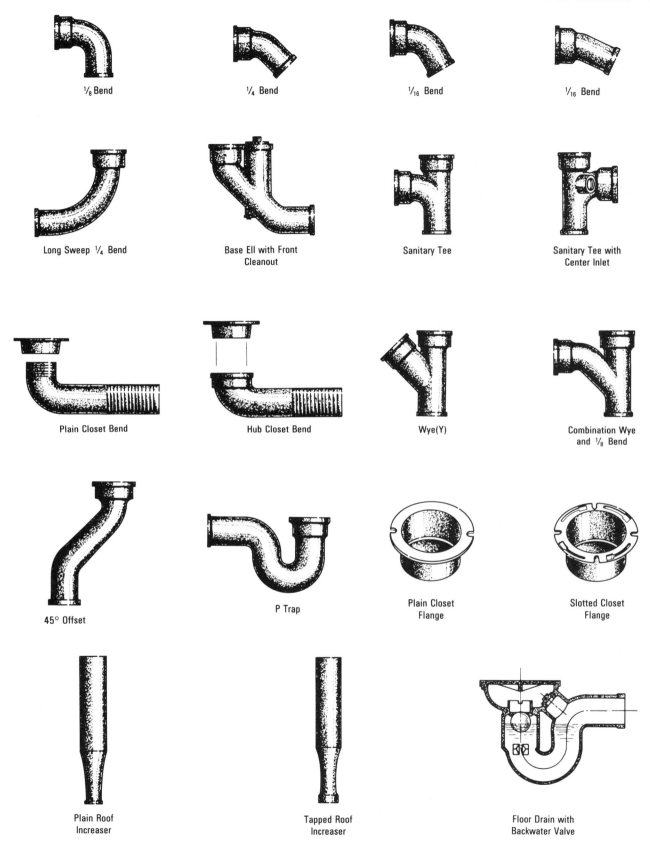

¹⁄₈ Bend

¼ Bend

¹⁄₁₆ Bend

¹⁄₁₆ Bend

Long Sweep ¼ Bend

Base Ell with Front
Cleanout

Sanitary Tee

Sanitary Tee with
Center Inlet

Plain Closet Bend

Hub Closet Bend

Wye(Y)

Combination Wye
and ¹⁄₈ Bend

45° Offset

P Trap

Plain Closet
Flange

Slotted Closet
Flange

Plain Roof
Increaser

Tapped Roof
Increaser

Floor Drain with
Backwater Valve

Fig. 18-2. Various bends and fittings.

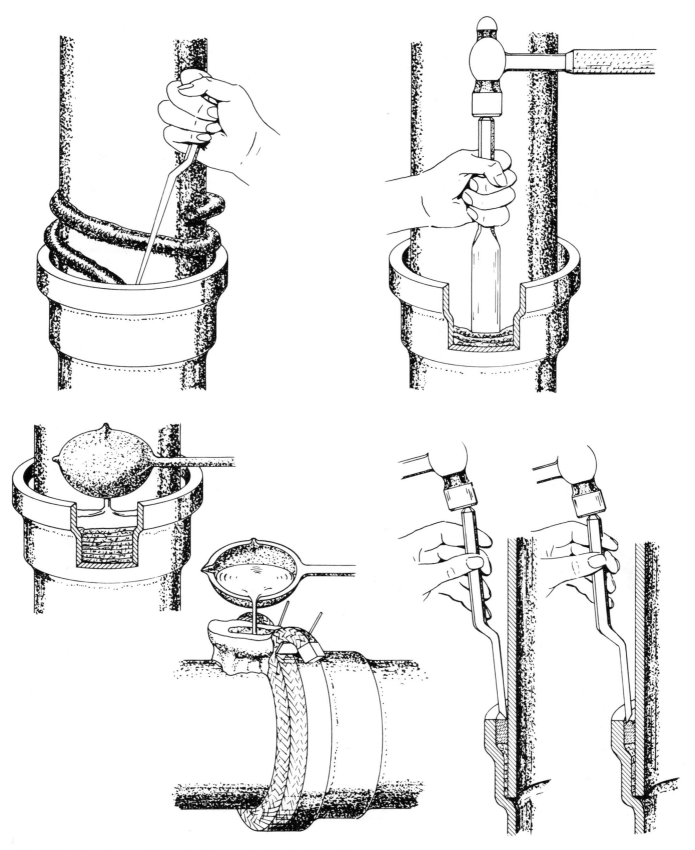

Fig. 18-3. Procedure used to pack oakum and molten lead into joints.

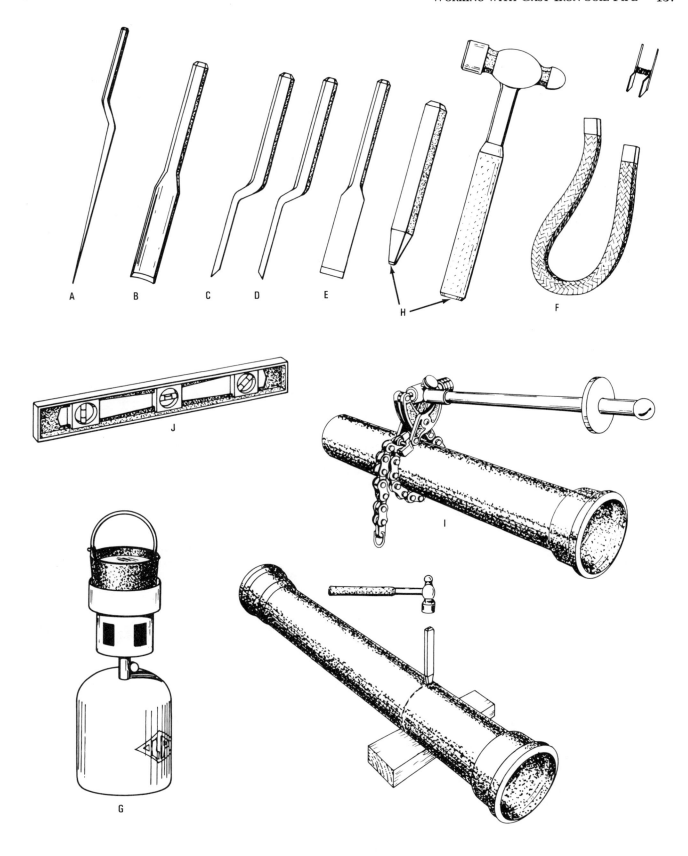

Fig. 18-4. Tools used when working with cast-iron soil pipe.

(B) Packing iron—used to pack the oakum tight in the joint.

(C) Inside caulking iron—used to tighten the lead joint against the pipe.

(D) Outside caulking iron—used to tighten the lead joint against the hub.

(E) Gate chisel—used to cut off the "gate" formed when pouring a horizontal joint.

(F) Joint runner—a braided asbestos rope that is placed around a horizontal joint to hold the molten lead until the lead has solidified. The joint runner is held in place by a clamp. A new joint runner should be soaked in light oil until the runner is soft and pliable.

(G) Propane tank, burner, hood, and lead pot—used to melt lead.

(H) Cold chisel and hammer—can be used to cut soil pipe. Use soapstone or chalk to mark on pipe where the cut is to be made. Place a 2 × 4 under the mark for the cut, use the chisel and hammer, rolling the pipe at the same time, and cut a groove around the pipe, on the cut mark. Tap the chisel lightly the first time around the pipe; the second and third times around the hammer blows can be harder; the pipe should break on the cut mark the third time around.

(I) Chain-type soil pipe cutter—the chain setting is adjustable for various sizes of soil pipe and the ratchet action makes cutting of soil pipe a quick and easy job.

(J) Level—used to establish the proper slope or grade of a pipe run. The grade, or fall, should be not less than $\frac{1}{8}$″ per foot; $\frac{1}{4}$″ per foot is the best rate of fall. When using a level, the bubble always goes to the high side.

CHAPTER 19

Septic Tanks and Disposal Fields

Septic tanks and disposal fields are used by many residents of suburban or rural areas that have no sewage treatment facilities. Precast concrete septic tanks, properly sized and correctly installed (and built to comply with local codes and health department regulations) will give many years of satisfactory service.

Percolation Tests

For a septic tank system to work properly, the effluent (discharge) from the tank must be disposed of. The best way to do this is to install a disposal field. The rate of absorption of water into the ground at a given point is the basis for determining the size of the disposal field. Percolation tests establish the rate of absorption.

The tests are made by digging three holes in the area where the disposal field is to be located. These holes should be from 2½ to 3' in depth. This is the correct depth for a disposal field. Fill the holes with water. When the water has been absorbed into the ground, begin the test. (This saturates the ground and simulates the actual working conditions of the soil when the disposal field is installed.)

Water should be poured into the holes to a depth of 10" from the bottom of the hole—do not add any water during the test. Lay a board across the top of the hole and measure the distance from the bottom of the board to the top of the water. Record this distance and the time. (Local health authorities usually have forms for this purpose.) After one hour, again measure the distance from the board to the top of the water. The difference between the first measurement and the second one is the inches-per-hour absorption rate. Continue the test by recording the distance down to the water level until the test has been conducted for five consecutive hours, or until the water has been absorbed. If the water is absorbed before the end of the five-hour test period, the test will be ended when the water has been absorbed. Generally speaking, the last hourly rate of absorption during the test is the figure to use in computing the size of the disposal field. If the rate of absorption is less than 1" in 60 minutes, a disposal field will probably not perform satisfactorily in that area.

SEPTIC TANKS

Local regulations will probably govern the minimum size of the septic tank. One commonly used rule is to figure that the storage capacity of the tank should equal the number of gallons of sewage entering the tank in a 24-hour period. At the rate of 100 gal. per person per day, a septic tank for a four-person household (or a two-bedroom home) should have a minimum capacity of 400 gal. storage. If a garbage disposer is used, the capacity should be increased 50 percent. (From 400 to 600 gal. for a four-person

household.) In actual practice, the minimum size of any septic tank should be 1000 gal.

Septic tanks function by a combination of bacterial action and gases. Solids entering the tank drop to the bottom. Bacteria and gases cause decomposition to take place, breaking down the solids into liquids, and in the course of this process the indissoluble solids, or sludge, settle to the bottom of the tank. Decomposition in an active tank takes about 24 hours. The sludge builds up on the bottom of the tank so that periodically (perhaps once a year, or only once in ten years, depending on usage) the tank must be cleaned or pumped out. In the cleaning or pumping process, only the sludge on the bottom of the tank should be removed. The crust, formed on the water level at the top of the tank, should not be disturbed. The only exception to this rule is if the crust has become coated with grease and the bacterial action of the tank thus destroyed. If this should happen, the crust on top will have to be removed.

Bacterial action will begin again when the top is placed on the tank and the tank sealed. Special compounds can be purchased to hasten the resumption of bacterial action but these compounds are rarely, if ever, really needed.

The top of the septic tank should be located at a minimum depth of 12 in. below ground level. The actual depth will probably be somewhat greater due to the depth of the sewer entering the septic tank. The septic tank must also be located at least 75 ft. away from any well, and downhill so that all drainage is away from the well. Local regulations must be followed as to placement of the septic tank.

THE DISPOSAL FIELD

The disposal field must be sized using the rate of absorption established in the percolation tests. If the absorption rate is 1″ in 60 minutes, then the factor 2.35 times the number of gallons of sewage entering the septic tank per day will determine the number of square feet of trench bottom area needed. Using the recommended figure of 100 gal. per day per person, if the absorption rate is 1″ in 60 minutes and there are four persons in the household, the formula 2.35 times 400 = square feet of trench bottom needed for the disposal field. If the disposal field has five fingers (trenches), each trench would need 188 square feet of trench bottom. If the trench is 2½ ft. (30 in.) wide, then each trench would be 75 ft. long.

If the absorption rate established in the tests was 1″ in 10 minutes, then the factor 0.558 times 100 gal. per person per day would be used. Again using the base of four persons in the household, then 0.558 times 400 = square feet of trench bottom needed. With five fingers or trenches installed in the disposal field, each trench should have 45 square feet of trench bottom. If the trench is 2½ ft. wide, then each trench should be 18 ft. long. The figures used in these examples are from the Indiana State Board of Health recommendations. Local regulations in your area may vary somewhat. The rules and regulations governing sizes of disposal fields in other areas should be followed.

Disposal fields serve two purposes. The fingers of a disposal field provide storage for the discharge of a septic tank until the discharge can be absorbed by the earth. The fingers also serve as a further step in the purification of the discharge through the action of bacteria in the earth. Liquids discharged into the disposal field are disposed of in two ways. One is by evaporation into the air. Sunlight, heat, and capillary action draw the subsurface moisture to the surface in dry weather.

In periods of wet or extremely cold weather, the liquids must be absorbed into the earth. The discharge from the septic tank should go into a siphon chamber and be siphoned into a distribution box. The distribution box should direct the discharges so that approximately the same amount of liquid enters each trench.

A siphon (also called a "dosing" siphon) is desirable for this reason:

The siphon does not operate until the water level in the siphon chamber reaches a predetermined point. When this level is reached, the siphonage action starts and a given number of gallons is discharged into the disposal field. The sudden rush of this liquid into the distribution box, and then into the separate fingers of the disposal system, insures that each finger will receive an equal share of the discharge. The discharge thus received by the fingers will have time to be absorbed before another siphon action occurs. A siphon is not essential to the operation of a septic tank and disposal field; however, a siphon will improve the efficiency of the system. The lateral distance between trenches in a disposal field will be governed by ground conditions and local regulations.

The construction of the disposal field will vary, due to soil conditions; basically, a typical trench will be as shown in Fig. 19-1. A layer of gravel is placed in the bottom of the trench, field tile or perforated drain tile is then laid on the gravel. If field tile (farm tile) is used, the tiles should be spaced ¼ in. apart and should slope away from the distribution box at the rate of 4 in. fall per 100 ft. The space between the tile should be covered by a strip of heavy, asphalt-coated building paper or by a strip of asphalt roofing shingle.

The trench should then be filled to within 6 in. of

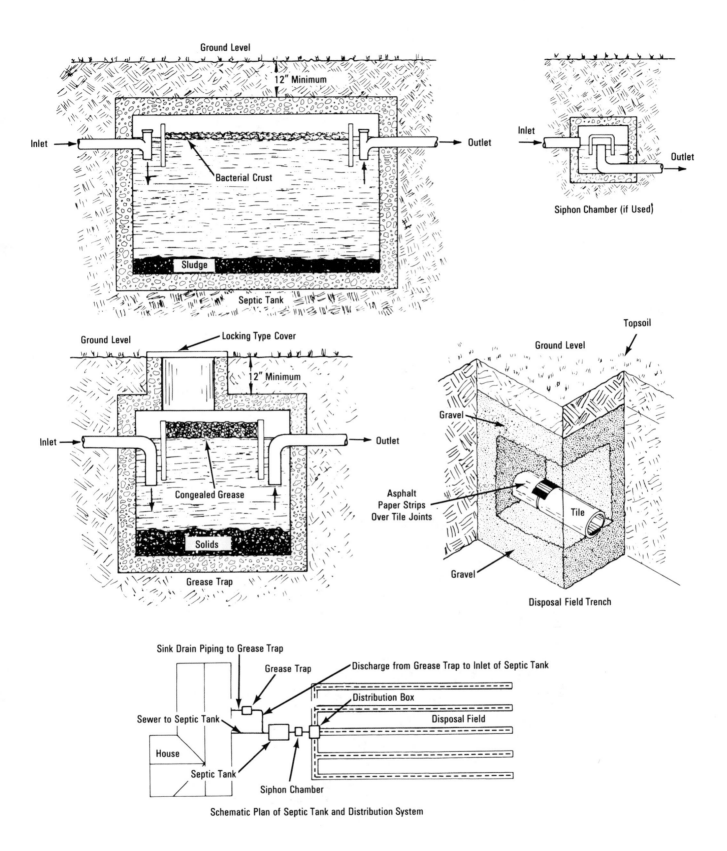

Fig. 19-1. The construction of a typical septic tank.

ground level with gravel. Topsoil should then be added to bring the finished trench to ground level. Since some settling of the topsoil will occur, it is wise to mound the earth slightly over the trenches to allow for settling.

Grease Trap

Grease, which is present in dishwashing water, can destroy the action of a septic tank, and if allowed to get into the disposal field will coat the surface of the earth in the disposal field and prevent absorption. Grease is present in sewage; the septic tank can dispose of this. If the grease from the kitchen can be trapped before it reaches the septic tank, the life of the septic tank and the disposal field will be greatly prolonged.

A grease trap must be large enough so that the incoming hot water will be cooled off as soon as it reaches the grease trap. A 400-gal. septic tank is ideal

for use as a grease trap. The water present in this size tank will always be relatively cool, and the grease, present in the dishwashing water, will congeal and rise to the top of the tank. The inlet side and the discharge side of the tank have baffles, or fittings turned down.

Thus, the water entering and leaving the tank is trapped. The grease will congeal and float to the top of the tank and relatively clear water will flow out of the grease trap and into the septic tank. Drain cleaning compounds that contain lye (sodium hydroxide) should never be used with a septic tank installation. The drainage piping from the kitchen sink will not be connected to the drainage system of the house when a grease trap is used. The kitchen sink drainage should be piped out separately and the grease trap located as near as possible to the point where the sink drain line exits from the house.

The top of the grease trap should be within 12″ of the ground level. A manhole can be extended up to ground level, with a lightweight, locking-type man-

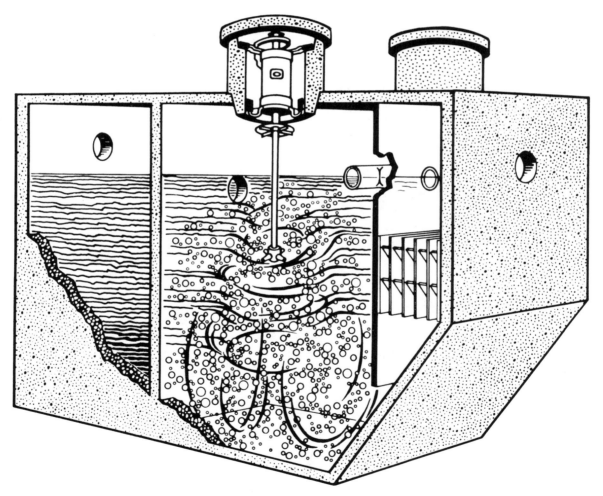

Courtesy Jet Aeration Company

Fig. 19-2. *Jet* aeration sewage tank.

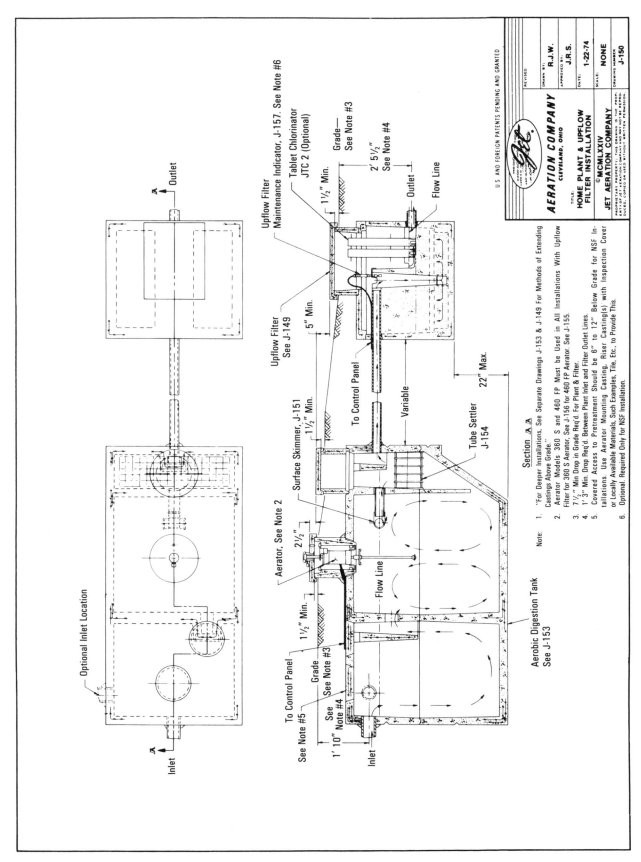

Optional Inlet Location

Outlet

Upflow Filter
Maintenance Indicator, J-157. See Note #6

Tablet Chlorinator
JTC 2 (Optional)

Grade—
See Note #3

1½" Min.

2' 5½"
See Note #4

Outlet

Flow Line

Upflow Filter
See J-149

5" Min.

To Control Panel

22" Max.

Variable

Surface Skimmer, J-151

1½" Min.

Aerator, See Note 2

2½"

Tube Settler
J-154

Flow Line

Aerobic Digestion Tank
See J-153

To Control Panel

See Note #5

Grade
See Note #3

1½" Min.

See
Note #4

1' 10"

Inlet

Section 𝔸·𝔸

Note: 1. "For Deeper Installations, See Separate Drawings J-153 & J-149 For Methods of Extending Castings Above Grade."

2. Aerator Models 360 S and 460 FP Must be Used in All Installations With Upflow Filter for 360 S Aerator, See J-156 for 460 FP Aerator. See J-155.

3. 7½" Min Drop in Grade Req'd. For Plant & Filter.

4. 1' 3" Min. Drop Req'd. Between Plant Inlet and Filter Outlet Lines.

5. Covered Access to Pretreatment Should be 6" to 12" Below Grade for NSF Installations. Use Aerator Mounting Casting, Riser Casting(s) with Inspection Cover or Locally Available Materials, Such Examples, Tile, Etc. to Provide This.

6. Optional. Required Only for NSF Installation.

Fig. 19-3. Dimensions of the *Jet* aeration tank.

hole cover; this will provide easy access for skimming off the grease. A periodic check should be made and the grease skimmed off when it reaches a depth of 3 or 4 in.

The tile in the disposal field will vary in depth according to local conditions. The septic tank supplier, if unfamiliar with a siphon and siphon chamber, can get this information. If a siphon is not used, the outlet of the septic tank is piped directly to the distribution box. The siphon outlet, if a siphon is used, must be lower than the inlet to the siphon chamber.

AERATED SEWAGE TREATMENT PLANTS

As we mentioned earlier, septic tanks and disposal fields have until recently been the only method by which residents of areas not served by central sewage systems could dispose of their sewage. The central sewage systems use a combination of bacterial action and aeration to break down sewage. If the plant is operating properly, the discharge or effluent is a clear, odorless liquid that can be piped to any running stream, discharged into a storm sewer, or routed through a disposal field where, by a combination of absorption and evaporation, it is finally disposed of. A septic tank and disposal field, or finger system, will give satisfactory service, provided that one essential condition exists at the septic tank and disposal field or finger system location. The ground must be able to absorb water at the disposal field or finger system depth, at a rate equal to 100 gal. per person per day, and it must be able to do this the year around. When this condition exists, the disposal field will work very well for a period of time.

When the soil conditions in an area are such that a septic tank installation will not work properly, the installation of an aeration-type of sewage treatment plant should be strongly considered. Small, one-household-size, aeration-type sewage treatment plants using the same methods as the large central plants are now available. These plants are very efficient; when operating properly the effluent discharge is clear, odorless, and can be chlorinated if necessary to meet health department standards. Local health departments often insist on aeration-type plants instead of septic tanks, especially where the water table is high or where the soil has shown poor percolation. The treat-

ment process, called extended aeration, is a speeded-up version of what happens in nature when a river tumbles through rapids and over waterfalls, purifying itself by capturing oxygen.

Does this type plant really work? The plant in Fig. 19-2 has been proven in the field and tested by private and government organizations, and has been accepted by the National Sanitation Foundation.

How does it actually work? This plant employs a biochemical action in which aerobic bacteria, using the oxygen solution, break down and oxidize household sewage. The three separate compartments in the plant each perform a specific function in the total purification process. The primary treatment compartment receives the household sewage and holds it long enough to allow solid matter to settle to the sludge layer at the bottom of the tank. Here an aerobic bacterial action continuously breaks down the sewage solids, both physically and biochemically, pretreating and conditioning them for passage into the second, or aeration, chamber.

In the aeration chamber the finely divided pretreated sewage from the primary treatment compartment is mixed with activated sludge and aerated. The *Jet* aerator circulates and mixes the entire content while injecting ample air to meet the oxygen demand of the aerobic digestion process. The *Jet* control panel, furnished as part of the plant package, is set to cycle automatically the running time of the aerator each day. The final phase of the operation takes place in the settling/clarifying compartment. In this compartment, a tube settler eliminates currents and enhances the settling of any remaining suspended material, which is returned, via the tank's sloping end wall, to the aeration chamber for further treatment. A nonmechanical surface skimmer, operated by tank roll, continuously skims any floating material from the surface of the settling compartment and returns it to the aeration compartment. The odorless clarified liquid flows into the final discharge line through the baffled outlet. Effluent disposal must conform to the requirements of the health authorities having jurisdiction. Normally, the highly treated *Jet* effluent eliminates the need for leaching fields or finger systems, or subsurface filters. In most areas, *Jet* effluent is discharged to a flowing stream, a storm sewer, or any well-defined line of drainage. An upflow filter and a chlorinator can be installed in this treatment plant if local conditions require it. Fig. 19-3 gives dimension of the *Jet* aeration sewage system.

CHAPTER 20

Private Water Systems

A good well or other source of water is a necessity for residents of suburban or rural areas not served by a water utility. The water furnished by a public or private utility is subject to inspections by health authorities. Water obtained from any other source, such as wells, springs, lakes, etc., should be tested for purity before it is used for human consumption. State and/or local boards of health or private laboratories will furnish sterile containers for water samples and test the water for purity. If the water is contaminated, filters, chlorinators, or iodine feeders can be installed in the private water system to make the water safe for human consumption.

Shallow well pumps can be used when the water does not have to be lifted more than 20'. Reciprocating-type (or piston-type) pumps and shallow-well-ejector-type (jet-type) pumps are best suited for shallow well usage. Where water must be lifted more than 20 ft., a deep well pump should be used. "Convertible-type" jet pumps can be used in either shallow wells or deep wells. The ejector can be mounted on the pump for use in shallow wells, or installed in the well casing for use in deep wells. Submersible-type pumps can be used on either shallow or deep wells. Submersible pumps do not need protection from freezing conditions; they are installed in a well below the water level.

PRESSURE/STORAGE TANKS

For proper operation, the water level in a pressure tank should be on the basis of two-thirds water, one-third air. With the top third of the tank filled with air, water is pumped into the tank, the air is compressed until the top (or high) setting of the pressure switch is reached, usually 40 lb. The pressure switch then turns off the pump. Water can then be used from the tank. As the water is pushed out of the tank, the pressure drops until the low setting of the pressure switch is reached (usually 20 lb.) and the pump starts, beginning the cycle over again. Occasionally, due to the failure of an air volume control, or a leak in the pressure tank, the air cushion (the top third of the tank) is lost or diminished.

If there is only a very small area of air at the top of the tank, the pressure will drop from 40 lb. to 20 lb. and immediately upon opening a faucet or valve, the pump will start. If there is no air at the top of the tank, the pump will operate continuously because it will be unable to build up enough pressure to cause the pressure switch to turn it off. When these conditions occur, the tank is "water logged." A water logged tank should be corrected as soon as possible; the frequent on-off operation of the pressure switch will result in burned

145

contacts on the switch and possible damage to the pump motor.

The air volume controls should be checked for proper operation; if a float-operated air volume control is used, check to see that the float is not sticking and that the float ball has not lost its buoyancy. *Schrader* air valves, or snifter valves, should be checked to see that they open and close freely. The pressure tank can be checked for leakages at any openings when pressure is built up by covering the openings with soap suds; any air leakage will then show up as soap bubbles.

Correct a waterlogged tank condition by turning off the electric power to the pump. There should be a drain valve at the bottom of the tank. Open this valve and drain the tank. In order to drain the tank properly, the air volume control, or a plug at or near the top of the tank, should be removed. Water draining from the bottom of the tank will cause a partial vacuum to be formed in the tank; removing the air volume control or the plug will relieve this vacuum.

The tank must be completely drained; the tank will then be filled with air at atmospheric pressure (14.7 lb. at sea level.) When the tank has been completely drained, replace the plug or the air volume control. These fittings must be tight in the tank with no air leakage when the system is in operation. Close the drain valve and turn on the electric power to the pump. If the pump lost its prime in this operation, it may have to be primed at this point. When the pump starts pumping water, the air in the tank will be compressed and forced into the top third of the tank. If there is no air leak in the tank, and if the air volume controls are working correctly, this ratio of one-third air and two-thirds water will be maintained.

The operation instructions included with the pump will show the location of the priming plug. The pressure tank can also become air-bound. This condition can occur when there is a leak in the suction pipe to the well, when the water drops below the end of the suction pipe, when there is a leak in the tubing between the pump and the air volume control on the tank, or by a leak in the air volume control device.

Air Volume Controls

There are several methods of maintaining the proper air volume in a storage tank. Snifter valves, such as a *Schrader* air valve, will draw in air when the pump runs. Diaphragm-type air volume controls, such as the

Courtesy Amtrol, Inc.

Fig. 20-1. A typical *Well-x-Trol* tank.

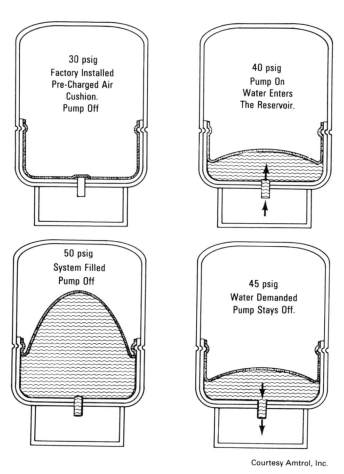

Courtesy Amtrol, Inc.

Fig. 20-2. *Well-x-Trol* sequence of operation.

Brady air volume control, will deliver a charge of air each time the pump turns on. Float-type air volume controls regulate the volume of air in the tank by opening to admit air into the tank when the water level raises the float, and expelling air when the water level in the tank causes the float to drop. Float-type controls have the disadvantage of sticking in one position, or the float can leak and lose its buoyancy.

PRESSURE TANKS

Pressure tanks that use a permanently sealed-in air charge, similar to the tank shown in Fig. 20-1, can be used instead of the old standard type. There are many advantages in the use of a permanently sealed type. Installation is much simpler, since there is only one opening in the tank. Waterlogging is eliminated by the sealed-in air cushion. The elimination of waterlogging keeps pump starts to a minimum, thereby adding to the life of the pump and the control switch. The water reservoir in the tank is corrosion-free.

Fig. 20-2 shows the sequence of operation of this type of tank, and also shows how waterlogging is pre-vented. Fig. 20-2(A) shows the tank installed; the pump is off. In (B), the pump is on, and water enters the reservoir, compressing the air cushion and raising the air cushion pressure. In (C), the system is filled. Pressure at 50 psig is applied to the water system, and the pump shuts off. In (D), water is used, the pressure in the air cushion drops until it reaches the cut-in point, and the pump starts, repeating the cycle.

Well-x-Trol tanks are made in sizes ranging from 2.0 gal. to 410.0 gal. Multiple installations can be made, as shown in Fig. 20-3, when greater capacity is required. Typical single installations of *Well-x-Trol* tanks are shown in Fig. 20-4.

PIPING CONNECTIONS AT THE WELL CASING

A sanitary well seal is a device (usually using bolts around the outside top side) to compress two plates together and expand a rubber seal against the inside of a well casing. The seal will have two holes in the top for use with an ejector (jet) installed in the casing, or one hole in the top for use with a shallow well jet or

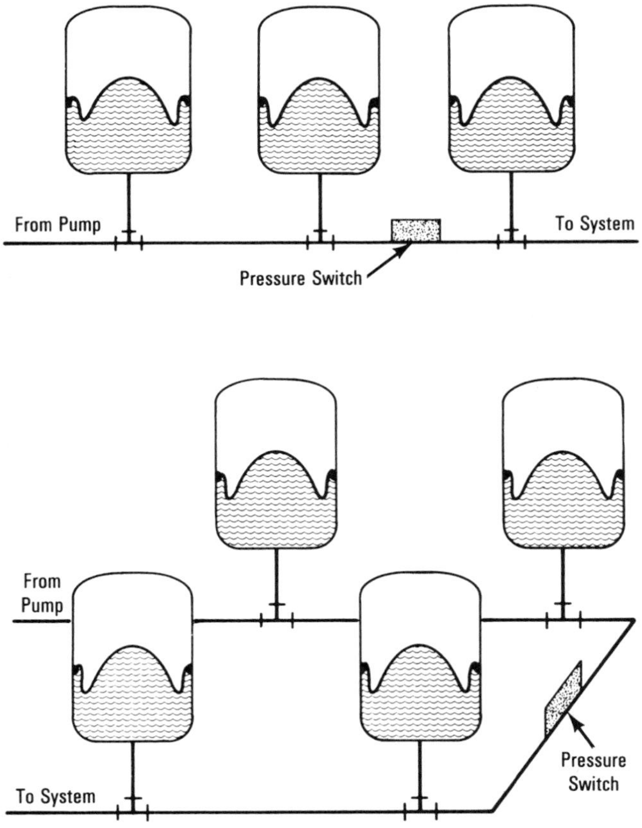

Fig. 20-3. Multiple installation for greater capacity.

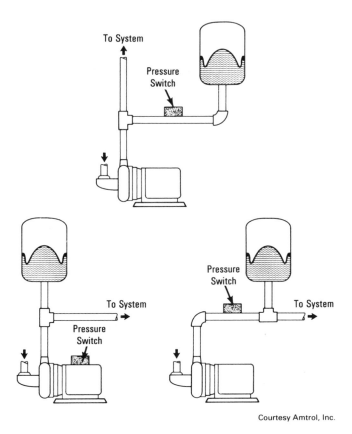

Courtesy Amtrol, Inc.

Fig. 20-4. Single installation of *Well-x-Trol* tank.

submersible pump. A sanitary well seal should always be installed where the piping enters the well casing to prevent contamination from surface water of the well. The water systems of most of our cities today are supplied with water obtained from reservoirs. This "raw" water is processed in filtration plants and chemically treated. Chlorine, aluminum sulphate, and activated carbon are added to complete the purification process.

Many rural residents, living on lakeside property, now have their own individual water systems based on the same theory as the large city water systems. If the lake water has not been polluted by discharges from septic tanks or run-off from stock feeding and grazing areas, this type of water system is very satisfactory. The first step in construction of this type system is to dig a hole near the edge of the lake. The hole should be 8' to 10' in diameter and approximately 10' deep. A 12"

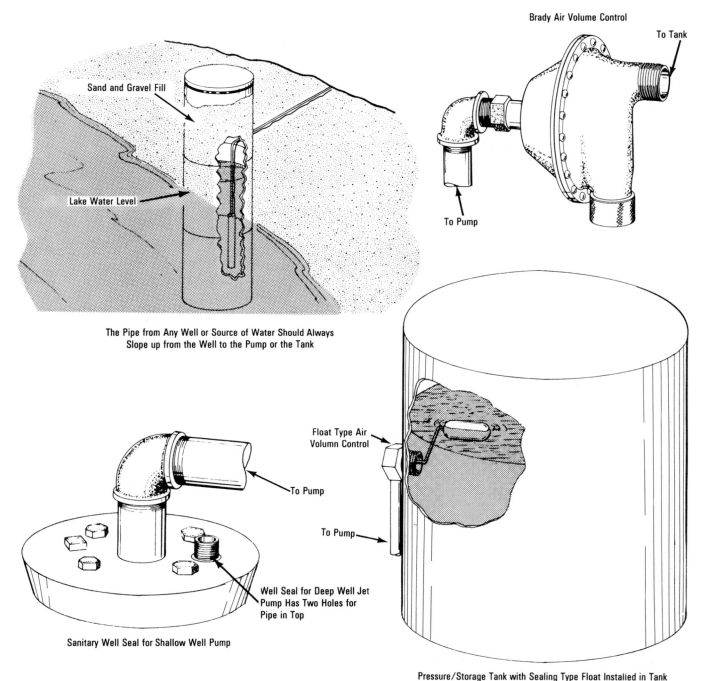

Brady Air Volume Control

To Tank

To Pump

Sand and Gravel Fill

Lake Water Level

The Pipe from Any Well or Source of Water Should Always Slope up from the Well to the Pump or the Tank

Float Type Air Volumn Control

To Pump

To Pump

To Pump

Well Seal for Deep Well Jet Pump Has Two Holes for Pipe in Top

Sanitary Well Seal for Shallow Well Pump

Pressure/Storage Tank with Sealing Type Float Installed in Tank

Fig. 20-5. Various inner working parts of a well water system.

layer of clean sand and small gravel is then placed in the bottom of the excavation. Concrete pipe 24″ to 30″ in diameter is then set down in the hole, extended up to 2′ above high water level of the lake. A mixture of clean sand and small gravel is then filled in around the concrete pipe. A 3-wire/220 volt submersible-type pump is the best type of pump to use with this type system. The pump, with 1″ discharge pipe, is set down in the well and the discharge pipe and the electric wiring to the pump are taken out of the side of the well (Fig. 20-5). The trench for the discharge piping and the electric wiring for the pump must be deep enough

Convertible Type Pump

Shallow Well Ejector

Convertible Ejector

Air-E-Tainer Float Type Tank

Multi-Stage Deep Well Pump

Submersible Pump

Iodizer

Tank Float Can Be Installed in Existing Tank to Prevent Waterlogging

Courtesy Flint & Walling, Inc.

Fig. 20-6. Various types of pumps used in well water systems.

to be below freezing level. A minimum depth of 54″ (4½′) is recommended; 1″ diameter plastic pipe, available in 100′ rolls and made especially for this purpose, is excellent for use as the discharge piping from the pump to the tank location.

A final filtration system, including a charcoal filter and an iodine feeder, or a chlorinator, can be installed at the tank location. The top of the "well" must be sealed with a concrete top, set in mortar, to prevent surface contamination.

A well with a high iron content in the water will cause deposits of iron oxide to form in the piping, tanks, water heaters, and water closet tanks. Once these deposits have formed it is virtually impossible to remove them. Water pressure tanks, water heaters, and piping may have to be replaced. An iron filter should be installed in any piping system where the water has a high iron content. Filters can also be installed to remove the objectionable taste and odor of so-called "sulfur" water.

Appendix

It helps to know the name of the fitting you need. Illustrated in Fig. A-1 are all of the common steel pipe fittings. Elbows are available as straight pipe size and also as reducing elbows— $\frac{1}{2}''$ elbows or $\frac{1}{2}'' \times \frac{3}{8}''$, $\frac{1}{2}'' \times \frac{3}{8}''$, $\times \frac{1}{2}'' \times \frac{1}{4}''$, etc. Tees are also available either as straight pipe size or as reducing tees— $\frac{1}{2}'' \times \frac{1}{2}'' \times \frac{1}{2}''$ (straight pipe size) or $\frac{1}{2}'' \times \frac{3}{8}'' \times \frac{3}{8}''$ or $\frac{1}{2}'' \times \frac{3}{8}'' \times \frac{1}{2}''$ or $\frac{1}{2}'' \times \frac{1}{2}'' \times \frac{3}{8}''$, etc. A tee is always "read" as end/end/side; thus, a tee which is $\frac{1}{2}''$ on one end, $\frac{1}{2}''$ on the other end, and $\frac{3}{8}''$ on the side is a $\frac{1}{2}'' \times \frac{1}{2}'' \times \frac{3}{8}''$ tee.

Couplings are often confused with unions. Couplings join pieces of pipe together as do unions, but unions must be used when joining two pieces of pipe together between pipes or fittings which cannot be turned or moved.

Pipe nipples are short lengths of threaded pipe, from "close" or all-thread to 12" length. Street elbows, both 45° and 90°, are not recommended; the male (outside threaded) end is restricted and street elbows are difficult to grasp with a pipe wrench. A 90° elbow and close nipple or a 45° elbow and close nipple will work much better than street 90° or 45° elbows.

Traps on fixtures are not there for the purpose of stopping or catching any object entering the drain of a fixture except as noted. Certain hospital-type fixtures, or fixtures used in connection with the making of casts, have a plaster-catching trap. The traps on the fixtures in the home are there for the purpose of preventing sewer gas from entering the home through the fixture drain piping.

RELIEF VALVES

Relief valves on water heaters will often, after several years of service, start leaking or will open and discharge a quantity of water before shutting off again. This usually is not due to a defective relief valve; many communities are outgrowing the old water mains installed many years ago and the water utilities are raising the pressures on the water mains in order to force more water through the now undersized mains as water usage increases. Consult your water utility if you have this problem; they can inform you on the correct pressure rating of a new relief valve.

CONDENSATION

Condensation forming on the outside of the water closet tank in the bathroom is a problem often experienced, especially in the winter season. When the warm moist air of the bathroom comes in contact with the cold surface of the water closet tank, the moisture in the air is condensed and forms as water droplets on the surface of the closet tank. If the water in the closet tank can be held at approximately the air temperature of the bathroom, no condensation will form. A thermostatic supply valve, which mixes hot and cold water to supply the water closet tank, can be obtained from your local plumber. Follow the installation instructions, and condensation on the closet tank will be eliminated.

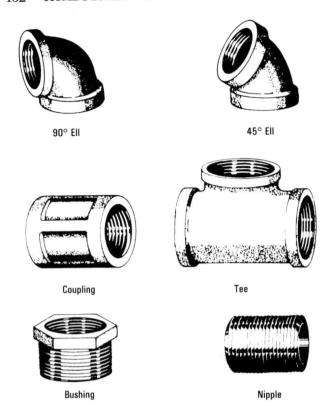

90° Ell   45° Ell

Coupling   Tee

Bushing   Nipple

Fig. A-1. Various pipe fittings.

THAWING FROZEN PIPING

If the water pipes freeze and are in an accessible location, they can be safely thawed by pouring hot water on the frozen pipe or wrapping rags dipped in hot water around the frozen areas. A heat lamp can also be used for thawing frozen pipes if the lamp is a safe distance away from combustible materials. An electric hair dryer aimed at the frozen areas will often thaw the pipe. An electric heating tape wrapped around the frozen areas will also thaw the pipe. If none of these methods do the job, your local plumber has special tools, electric thawing machinery and steamers, that *will* solve the problem.

An ounce of prevention is worth a pound of cure! If there is a possibility that water pipes may freeze, open the faucet (hot and cold sides) and let a small stream of water flow. Water moving through the piping will not freeze.

NOISE IN THE PIPING SYSTEM

A hammering or crackling noise in a water heater is caused by lime or other minerals that solidify on the bottom or sides of a water heater when the water is heated. This mineral coating is not watertight and

small droplets of water are trapped under this coating. When the burner comes on, the trapped water is heated and expands. The expansion of this water which is trapped under the coating causes the coating to flake off and in the process the rumbling, hammering, or crackling noise is produced.

It is a common occurrence for a hammering noise to be produced when a faucet is turned on. The noise is generally heard if the faucet is only slightly opened, with a very small stream of water flowing. The noise then will be a slow hammering noise, which will increase in volume and intensity as the faucet is opened wider. At wide open position, the noise usually disappears. This noise is caused by a loose bibb washer. Remove the stem from the faucet and tighten the bibb screw holding the bibb washer. A whistling-type noise is often heard in the water closet tank. This noise is caused when an old-fashioned-type ballcock, shown in Fig. 4-1, starts to shut off the water flow as the proper water level in the tank is reached. The rate of water flow through an old-fashioned ballcock varies. The water flows at a fast rate when the tank starts to refill, but as the float ball rises with the water level and in turn pinches off the water flow, a whistling noise is often produced. The chapter on repairing water closet tanks explains the operation and installation of a new type water fill valve. The installation of this new type of valve will eliminate this source of noise.

Noise in the pipe is also caused by the sudden closing of a valve. This causes a shock wave to travel through the water pipe, creating a hammer-blow type of noise. This noise is very common during usage of an automatic electric dishwasher or an automatic clothes washer. These appliances have electrically operated solenoid valves installed in the water pipe and when these valves close, they close very quickly, literally slamming closed and creating a shock wave in the water pipe. There is a device manufactured to cure this problem. This is a water hammer arrester, or shock absorber. The ½ in. I.P.S. (iron pipe size) shock absorber is large enough for the average clothes washer or dishwasher. Best results will be obtained if a shock absorber is installed in both the hot- and cold-water pipes. Shock absorbers are not expensive and are available through plumbing shops.

GENERAL INFORMATION

Every homeowner should know where the main shut-off valve is located on the water supply to the home. Knowing where this valve is located and how to operate it may prevent costly damage. Follow the

cold-water connection from the water heater back to the point where the water pipe enters the house. The valve should be located at or near the outside wall. The main valve controls all water supply to the house. If it is necessary to shut off only the hot-water supply to the house, the shut-off valve on the *cold*-water supply to the water heater will turn off the hot water.

An emergency repair clamp can be easily made by using one or two adjustable radiator or heater hose clamps and a small inner-tube patch or small piece of inner tubing. If the clamp (or clamps) is placed over the hole and tightened, it will hold pressure until a permanent repair can be made.

Water backing up through the basement floor drains is a problem in some areas. Most basement floor drains have a 2-in. tapping in the body of the drain, under the cover. A 2-in. backwater valve can be purchased at hardware or plumbing stores. The backwater valve is a fitting with a rubber seat on the bottom of the fitting and a metal or plastic float ball. The float ball will drop to permit the passage of water through the inlet side of the floor drain, but the ball will raise and seat itself against the rubber seat of the fitting, preventing water from backing up through the floor drain.

A *vent* when applied to a plumbing system is a pipe that provides a flow of air to and from the plumbing drainage piping. Proper venting is necessary to permit good drainage flow and prevent siphonage in a drainage system.

To eliminate the possibility of accidental scalding, the thermostat on a water heater should be set to deliver a maximum hot-water temperature of 120°F. The hot-water temperature should be checked using a thermometer at the water heater or at a hot-water faucet nearest to the heater.

Bathtubs are now being made with slip-resistant surfaces. A great many injuries are caused by slipping and falling in a bathtub. Bathtubs now in use, which do not have "slip-resistant" bottoms, can be made much safer by the application of non-slip tapes.

HOW TO FIGURE CAPACITIES OF ROUND TANKS

There are occasions when it is necessary or desirable to be able to compute the contents in gallons of round tanks, cisterns or wells. A tank 12 in. in diameter and 5 ft. in length will hold 29.3760 gal. of water. A well casing 4 in. in diameter with 10 ft. (120 in.) of water standing in the casing has 6.528 gal. of water standing in the casing.

Shown are the two old standard methods of figuring the contents of round tanks. Also shown are two short-

er and easier methods. The shorter methods are not only easier but there is less chance for error, since fewer steps are used in obtaining the results.

C = capacity in gallons
D = diameter
L = length

0.7854 = area of circle
231 = cubic inches in gallons
7.48 = gallons in cubic foot

The old standard methods:

1) When measurements are in inches:

$$\frac{D^2 \times 0.7854 \times L}{231} = \text{capacity in gallons}$$

2) When measurements are in feet:

$$D^2 \times .7854 \times L \times 7.48 = \text{capacity in gallons}$$

PROBLEM: A tank is 12 in. in diameter and 5 ft. in length. How many gallons will the tank hold?

Using method 1—

```
1)        12    2)         144    3)      113.0976
        × 12          × 0.7854            ×     60
        ----          -------          ----------
         144              576           6785.8560
                          720
                         1152
                         1008
                       --------
                       113.0976
```

```
4)          29.376
     231 | 6785.8560
          462
          ----
          2165
          2079
          ----
           868
           693
           ----
          1755
          1617
          ----
          1386
          1388      Answer: 29.3760 gallons
```

Note: Four steps are necessary in arriving at the answer.

Using method 2—

```
1)      1    2)   0.7854   3)   0.7854   4)      3.927
      × 1          ×   1          ×  5          × 7.48
      ---         -------        -------        ------
        1          0.7854        0.39270        3.1416
                                                15708
                                                27489
                                               -------
                                               29.37396
```

Answer: 29.373 gallons

Note: Four steps are necessary in arriving at the answer; also, the answer does not agree with answer shown on method 1. The difference in the answers is minor, however.

The shorter methods in figuring tank capacities are:

Using method 3—[C = D²(in.) × L(ft.) × 0.0408] when measurements are in feet and inches

```
1)      12       2)      144      3)        720
      × 12              ×  5             × 0.0408
       144               720              5760
                                        28800
                                        29.3760
```

Answer: 29.3760 gal.

Note: Three steps are necessary using this method.

Using method 4—[C = D² × L × 0.0034 = gal.] when measurements are in inches

```
1)      12       2)      144      3)        8640
      × 12              × 60             × .0034
       144              8640             34560
                                        25920
                                        29.3760
```

Note: Only three steps are necessary using this method; also methods 3 and 4 arrive at the same exact answer when carried to four decimal points.

PROBLEM: A well casing is 4 in. in diameter, 120 ft. deep, and has 10 ft. of water standing in the casing. How many gallons of water are in the casing?

Using method 4—

```
1)       4       2)      16       3)      1920
       × 4             × 120            × .0034
        16              320             7680
                         16             5760
                        1920            6.5280
```

Using method 3—

```
1)       4       2)      16       3)       160
       × 4             × 10             × .0408
        16             160             1280
                                       6400
                                       6.5280
```

STEM INDEX BY MANUFACTURER

MANUFACTURER	FITS O.E.M. NUMBER:	PART NO.
American Brass	118 Concealed Sink Faucet	FS1-1
American Brass	104, 1104, 105, 1105 Sink	FS1-2
American Brass	100 Sink	FS4-1
American Kitchens	5771-C, 5772-H, LF-110, 120, 210	FS3-1
American Kitchens	4271-H, 4272-C, LF-100-C, 200-E	FS5-3
American Standard	36536-02 Built-in Bath & Shwr.	FS10-3
American Standard	21674-02(LH), 21186-02(RH) B-892, B-872	FS10-23
American Standard	25509-02 Tract Line Bath	FS10-25
American Standard	19276-02, K-370, K-371 (¾"x½") Exp. Bath Fitting	FS11-3
American Standard	21698-02 Diverter Stem Transfer Valve Fitting	FS11-66
American Standard	(Re-Nu Barrel) 20336-08 RH, 20563-08 LH	FS1-3
American Standard	64804-07 RH, 50637-07 LH, C-530-560 Colony Lav. Fitting	FS1-23
American Standard	6079-04-H, 6080-04-C, R-4100-1-2-3	FS2-1
American Standard	853-14-R, 856-14-L, R-4100-1-2-3	FS2-2
American Standard	54156-04-H, 54157-04-C, Crosley CF-100-200, Schaible 922-26, SCH-1622, Youngstn-Mull. 5505, 60954-55	FS2-3
American Standard	72950-07-RH, 72951-07-LH	FS2-11
American Standard	7617-07-LH, 7616-07-RH, B-876-8	FS3-2
American Standard	711-17-LH, 712-27-RH, B-876-8	FS3-3
American Standard	25535-02-RH, 25536-02-LH, TL-30-37 Tract Line Lav. Fitting	FS3-17
American Standard	21658-02-LH, 21659-02-RH, B906-4211	FS4-4
American Standard	64703-07-RH, 50668-07-LH, Colony Trim	FS4-27
American Standard	21451-02-LH, 20731-02-RH, B-900-2, B-901	FS5-4
American Standard	21883-02, F-305, P-4100-S Lavatory	FS6-1
American Standard	19376-02, F-105, F-115 Lav., B-787 Lavatory, B-912S Sink	FS7-1
American Standard	21668-02-LH, 21195-02-RH, B-874	FS7-2
American Standard	21706-02-LH, 21705-02-RH, P-3905 Lavatory	FS7-4
American Standard	55815-04-RH, 55816-04-LH, R-4046-48	FS8-2
American Standard	55838-04-RH, 55839-04-LH	FS7-13
Briggs	22268 Stem, 22269 Bonnet, T-8832-33 Trim Line Lavatory	FS1-21
Briggs	22291 Stem, 22292 Bonnet, For T-9153-59 Deck Fitting	FS4-73
Briggs	22097-RH-C, 22098-LH-H, T-8802-03 Slant-Back	FS5-7
Briggs	22033-C-RH, 22034-H-LH, T-8805-15 Shelf Back Lavatory	FS5-8
Briggs	6612-H, 6613-C, T-8200, T-9205	FS7-5
Briggs	5857: T-8105-15-25, T-8205-10 Bath & Shower	FS10-6
Briggs	67: 1170, 1200, 1224, 1230 Bath Fittings	FS11-5
Central Brass	SU-357-K Model Lavatory, Bath, Laundry and Sink	FS1-5
Central Brass	SU-1855-47 (R or L), 21, 23, 26, 28, 21-S, 26-S, 47½, 48½	FS3-15
Central Brass	SU-1994-L, SU-1994-R, 70-S, 71-S, 76-S	FS7-6
Central Brass	SU-1548R, 7868, 8868, 9868 Diverters	FS10-7
Chicago Faucets	217-X-LH, 217-X-RH, 889	FS6-3
Crane	F12537 Magic-Close Faucet	FS3-6
Crane	FB-8035-H, FB-8034-C, All Dial-Eze Faucets	FS4-7

MANUFACTURER	FITS O.E.M. NUMBER:	PART NO.
Crane	F12536-H, F12535-C, All Dial-Eze Faucets	FS4-8
Crane	New Sleeve Unit (25A-25 ⅜), C-32180-85-86, C-32165-66-67 Lav., C-32279-81-82 Lav., C-32835 Sink, C-32746-S Wall Sink	FS4-24
Crane	Sleeve Unit (25A-10¾) C-31806 Telsa, C-32267 Securo Jr.	FS5-16
Crane	F13167-RH, F13168-LH, Hospital Stem	FS6-5
Crane	New Sleeve Unit (25A-14¾) Imp. Telsa Wall Mount Sink Faucet	FS6-13
Crane	FB-7674, C-32475-450 Panel Back Lav., Criterion & Rival Lav., C-32820-450 Sinks. Complete Assembly. Pre-War Trim	FS9-23
Crane	FB-1341 (Pre-War) Complete Assembly	FS10-17
Crane	FB-1341 (Pre-War) Criterion & Rival Concealed Bath and By-Pass Valves	FS10-60
Crane	FB-7673, No. 1 Line Lav. (Pre-War) Complete Assembly	FS11-15
Crane	FB-1077 Complete Assembly (Pre-War) No. 1 Line Bath Valves, ½" and ¾"	FS12-1
Crosley	CF-100-200	FS2-3
Dick Bros.	Fits Dick Bros. 2011-H, 2011-C, D-4008-10 Wall Faucet	FS3-13
Dick Bros.	Fits Dick Bros. 3057 Deck Faucet	FS4-60
Elkay		FS4-28
Eljer	5288-1-RH, 5288-2-LH, No. 4 Unit	FS1-22
Eljer	4788 No. 3 Unit	FS4-25
Eljer	2807, 9575R, 9576R	FS4-72
Eljer	5182	FS4-61
Eljer	4650	FS4-62
Eljer	5186	FS5-61
Eljer	4290, E-9560, E-9562-R	FS5-67
Eljer	2733, E-9340-R-41R Concealed Lavatory Fittings	FS6-65
Eljer	2788, E-9564-5-6	FS7-64
Eljer	3045 Bath Fittings	FS8-66
Eljer	5259-1-RH, 5259-2-LH, Tub Valve Assembly, No. 5 Unit	FS9-29
Empire Brass	611 Ledge Type	FS1-1
Empire Brass	800 Ledge Type	FS1-2
Gerber	RPA-28-1, 29-1	FS1-6
Gerber	29-1: 250	FS1-61
Gerber	260, 225-40, 53, RPA-28-2, 29-2	FS2-4
Gerber	29-2, 260, 22540, 53	FS2-61
Gerber	393-1: 365, 368 Ledge Type Sink Faucets Since '61	FS4-12
Gerber	7-2 Tub Stem	FS10-8
Gerber	7-2, 12-1, 98-672	FS10-65
Gerber	RPA-13-1, 37-1, 98-722	FS11-64
Gopher-St. Paul	(See Union Brass)	
Harcraft Brass	3247, 10-100 through 10-171, 12-300 through 12-302	FS1-13
Harcraft Brass	99-3128-H, 99-3129-C, A-50, A-51-A, A-52-A, A-160, A-180, A-190, A-540, A-541 Swing Spouts, Center Sets, Exposed Ledge Types	FS1-14
Harcraft Brass	3140: Valve Stem for Diverters, Shower and Tub Fillers prior to '60	FS10-18
Harcraft Brass	3141: Diverter Stem for Diverters & Tub Fillers prior to '60	FS10-19

STEM INDEX BY MANUFACTURER

MANUFACTURER	FITS O.E.M. NUMBER:	PART NO.
Indiana Brass	631-C & D	FS1-7
Indiana Brass	552-C, w/582-D Gland Nut	FS8-61
Kohler	32462 RH, 39717 LH, (Renew Barrel)	FS1-8
Kohler	32473 (Renew Barrel) No Thread	FS1-62
Kohler	34070-C, 34071-H, K-8610-A, K-8005A-06-09	FS3-7
Kohler	31871-H, 31872-C, K-8686, K-8692, K-8655, K-8660	FS4-14
Kohler	20655-H, 20656-C, Valve Unit, Aquaric	FS4-63
Kohler	32491, K-8178, Single Lavatory Fitting	FS5-11
Kohler	22932-H, 22917-C Constellation, Galaxy, Triton, Aquaric	FS5-22
Kohler	31584-C, 31585-H, Hampton, Taughton, Marston, Grammercy, Strand; K-8100-15-32-33	FS7-10
Kohler	31591-H, 31592-C, K-8634-36, K-8638, K-8650	FS8-8
Kohler	37645-C (Cold), 37646-H (Hot) Concealed Lavatory	FS9-7
Kohler	20654 Valvet Unit Aquaric Bath & Shower	FS10-24
Kohler	20242, K-7030-32, K-7100, K-7210, K-7240-42-45-47 Bath & Shower Valves	FS11-10
Milwaukee	5012-H, 5012-C, K-4009 Lavatory Center Set	FS3-14
Noland (Sayco)	45D-12 for 280 Fitting	FS1-11
Price-Pfister	03205-01	FS3-63
Price-Pfister	635-645	FS1-9
Price-Pfister	03156-01-H, 03156-02-C	FS2-6
Price-Pfister	3147, 460-61 Single Lavatory	FS2-7
Price-Pfister	03175-01-00Z, 43-010 through 43-124 Fittings	FS2-14
Price-Pfister	03206-01	FS4-70
Price-Pfister	03220-01	FS4-71
Price-Pfister	03151-01-RH, 03151-02-LH	FS4-15
Price-Pfister	03155-01-RH, 03155-02-LH (Fits 703 Ledge Faucet)	FS6-7
Price-Plister	3108, 10 & 12 DLH Series Crown Imperial Bath	FS8-9
Price-Plister	3109	FS8-10
Price-Pfister	3110, 10, 12, 50, 60 Bath & Shower	FS10-10
Price-Pfister	03172-01-00-A	FS10-11
Price-Pfister	3111	FS10-68
Repcal Brass	14147 RH, 14148 LH, Old No. 16-5	FS1-10
Repcal Brass	FB-9173 RH, FB-9174 LH, Also 16-3A	FS1-19
Repcal Brass	F14143H, F14145C, 13-511, 205 Wall Sink, 123 Laundry	FS4-16
Repcal Brass	515-L-5	FS9-9
Repcal Brass	1112-5A, B-625P, B-626P, B-1137-P Bath Fittings	FS9-10
Repcal Brass	F14153	FS10-13
Repcal Brass	110-5	FS11-65
Republic Brass	(See Briggs)	
Richmond	18 Serration Broach (New Strle	FS6-63
Richmond		FS5-18
Savoy Brass	A-17-H, A-17-C, 18 Serration Broach (New Style)	FS6-63
Sayco (Noland)	45-D-12 Stem, 57-D-13 Bonnet For 280 Fitting	FS1-11
Sayco (Noland)	45-D-12 Stem for 280 Fitting	FS1-63
Sayco	255-R, 255-L, 1100 Sink Faucet	FS6-10
Sayco	214-R (Stem), 572-R (Bonnet)	FS8-64
Sayco	LOS-1: 206, 208, 308, 311, 406 Bath & Shower Valves (New Style)	FS9-13
Sayco	1-1116-R: 308, A-308-C, 208, 206, 20811 Bath & Shower (New Style)	FS9-64
Sayco	LOS-1D: Diverter Stem	FS9-65
Sears-Roebuck	Sears, Universal-Rundle Lav.	FS1-64
Sears-Roebuck	Sears-Homart	FS2-10

MANUFACTURER	FITS O.E.M. NUMBER:	PART NO.
Sears-Roebuck	Sears-Milwaukee 5077RH, 5078-LH	FS2-65
Sears-Roebuck	Sears, Elkay, Universal-Rundle: P-1077, 32008-L	FS4-28
Sears-Roebuck	Sears, Homart, Universal-Rundle	FS4-66
Sears-Roebuck	Sears, Homart, Universal-Rundle	FS5-64
Schaible	56628-H, 56629-C, (Old) 1932, 1936, 1941	FS6-11
Schaible	55838-04-H, 55839-04-C, New 932	FS7-13
Schaible	Schaible: 922-26, 1622	FS2-3
Schaible	957	FS5-5
Schaible	6618-H, 6619-C, 936	FS8-12
Speakman	3-336-C, 3-337-H, 3-291 (G-3-180) H, 3-292 (G-3-181) C, SK-561	FS2-9
Speakman	G3-158H, G3-159C, S-4760-61, S-4770-71, S-4700	FS3-9
Speakman	G3-140H, G3-139C, S-4700	FS5-14
Speakman	3-250H, 3-251C, S-4095, Diamond Lavatory	FS5-18
Speakman	G3-122 Bath Fitting	FS10-15
Speakman	3-258 Lavatory Fitting Stem	FS10-63
Sterling Faucet	99S-8079-H, 99S-8080-C, 20-310 Series Lavatory Fittings	FS1-17
Sterling Faucet	99S-8109-H, 99S-8110-C	FS1-18
Sterling Faucet	99S-8153-H, 99S-8154-C	FS2-12
Sterling Faucet	99S-8131-H, 99S-8132-C	FS2-66
Sterling Faucet	99S-8125-H, 99S-8126-C	FS3-19
Sterling Faucet	99S-8062-H, 99S-8063-C, 19-050, 19-060 Bath Fitting	FS3-20
Sterling Faucet	99S-8171-H, 99S-8172-C	FS3-21
Sterling Faucet	99S-0064-H, 99S-0294-C	FS4-21
Sterling Faucet	99S-0148-H, 99S-0648-C, S-200, S-1120-22	FS4-22
Sterling Faucet	99S-8021-H, 99S-8020-C, 20-10 (N-300) Lavatory Fitting	FS4-23
Sterling Faucet	99S-8128-H, 99S-8129-C	FS4-67
Sterling Faucet	99S-3193-H, 99S-3194-C	FS5-15
Sterling Faucet	99S-1080-H, 99S-1079-C	FS6-12
Sterling Faucet	99S-0235 (S-1100 Old)	FS7-15
Sterling Faucet	99S-8000 (Series 10-000) Tub Valve Assembly	FS9-19
Sterling Faucet	99S-8001 (Series 10-000) Diverter Assembly	FS9-20
Sterling Faucet	99S-0160-H, 99S-0316-C	FS9-30
Sterling Faucet	99S-8136	FS9-66
Sterling Faucet	99S-1023	FS9-67
Sterling Faucet	99S-1025	FS9-68
Sterling Faucet	99S-8005	FS10-16
Sterling Faucet	99S-0174: 10-200 (S-600 Series) Bath Fitting	FS10-64
Sterling Faucet	99S-8006	FS11-13
Streamway	118, 1118, 218, 1218 Concealed Ledge Type Sink Faucet	FS1-1
Streamway	108, 1108, 208, 1208 Sink Faucet; 104, 1105, 105, 1105 Lavatory; 103, 1103 Tub & Shower	FS1-2
Tracy		FS2-3
Union Brass	1837-A	FS1-12
Union Brass	3402, 1837-A, L-7370 (Also Gopher-St. Paul)	FS2-68
Union Brass	P-107-H, P-107-C, L-7335-41, L-7315-21 (Also Gopher-St. Paul)	FS3-11
Union Brass	1840-A-RH, 1840-A-LH, 30-2-3-4-5 Bath Fitting	FS7-17
Union Brass	P-106-H, P-106-C (Also Gopher-St. Paul)	FS8-67
Union Brass	Stem & Seat Holder	FS9-69
Union Brass	P-19: L-6300-06, L-6320-25, L-6404-05-06-08 Bath Fitting	FS9-70
Universal-Rundle	Sears, Elkay, Universal-Rundle P-1077, 32008-L	FS4-28
Universal-Rundle	Sears, Homart, Universal-Rundle	FS4-66
Universal-Rundle	Sears, Homart, Universal-Rundle	FS5-64
Universal-Rundle	R-39-H, R-39-C, 0-310-25	FS7-18
Youngstown-Mullins	5505, 60954-55	FS2-3
Youngstown-Mullins	5546-H, 5547-C	FS5-5

FAUCET STEMS

The faucet stems shown here are all actual size drawings, giving you the easiest, quickest solution to replacing old or worn faucet stems. All stems are grouped into 12 basic sizes, from No. 1 (smallest) to No. 12 (largest).

To identify your stem, simply place it on the drawings until an exact duplicate is found. This is your stem number. The text accompanying each drawing includes original manufacturer, OEM part no. and if hot or cold sides are different. If number includes "C" (FS1-9C), this is cold side. If number includes "H" (FS1-9H), this is hot side. If number includes "HC" (FS1-13HC), stem fits either side.

Stem packages are color-coded: Red for HOT side (H); Green for COLD side (C); Brown for Hot or Cold (HC). Order by COMPLETE part number.

When replacing stems, it is often advisable to replace old seats also. The replacement seat numbers required for each stem are shown (in most cases) along with each stem drawing.

BRASS BIBB SCREW: S-553
MONEL BIBB SCREW: S-1258

SEAT NO. S-2001
WASHER: W-145A (O-FLAT)

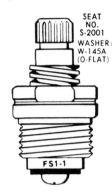

FS1-1C   FS1-1H
Fits AMER. BRASS, EMPIRE BRASS, STREAMWAY
118 Concealed Sink Faucet

BRASS BIBB SCREW: S-555
MONEL BIBB SCREW: S-1262

WASHER: W-145A (O-FLAT)

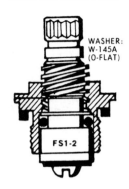

FS1-2C   FS1-2H
Fits AMER. BRASS, EMPIRE BRASS, STREAMWAY: 104, 1104, 105, 1105 Sink

---

FS1-3C   FS1-3H
Fits AMER. STANDARD (Re-Nu Barrel)
OEM 20336-08 RH. 20563-08 LH

BRASS BIBB SCREW: S-555
MONEL BIBB SCREW: S-1262

SEAT NO. S-1091-C
WASHER: W-146 (¼ S-FLAT)

FS1-5C   FS1-5H
Fits CENTRAL SU-357-K
Model Lav., Bath, Laundry and Sink

BRASS BIBB SCREW: S-552
MONEL BIBB SCREW: S-1263

SEAT NO. S-1091C
WASHER: W-148 (¼ FLAT)

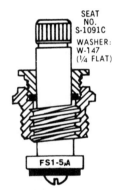

FS1-5AH   FS1-5AC
For all bath, lavatory, laundry and sink faucets. Interchanges with FS1-5.

BRASS BIBB SCREW: S-553
MONEL BIBB SCREW: S-1258

SEAT NO. S-2007
WASHER: W-148 (¼ L-FLAT)

FS1-6C   FS1-6H
Fits GERBER
OEM RPA-28-1, 29-1

BRASS BIBB SCREW: S-552
MONEL BIBB SCREW: S-1263

SEAT NO. S-2008
WASHER: W-147 (¼ FLAT)

FS1-7C   FS1-7H
Fits INDIANA BRASS
OEM 631-C & D

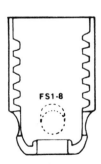

FS1-8C   FS1-8H
Fits KOHLER
OEM 32462 RH, OEM 39717 LH

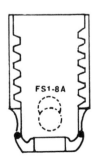

FS1-8AC   FS1-8AH
Same as OEM 32462 RH (FS1-8C)
Same as OEM 39717 LH (FS1-8H)
but with "O" Ring

BRASS BIBB SCREW: S-553
MONEL BIBB SCREW: S-1258

SEAT NO. S-2012
WASHER: W-145A (O-FLAT)

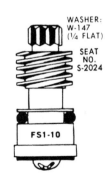

FS1-9C   FS1-9H
Fits PRICE-PFISTER
635-645

BRASS BIBB SCREW: S-553
MONEL BIBB SCREW: S-1258

WASHER: W-147 (¼ FLAT)
SEAT NO. S-2024

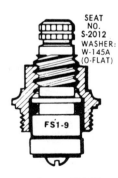

FS1-10C   FS1-10H
Fits REPCAL
OEM 14147 RH, OEM 14148 LH
Old No. 16-5

BRASS BIBB SCREW: S-555
MONEL BIBB SCREW: S-1262

SEAT NO. S-2012
WASHER: W-145 (00 FLAT)

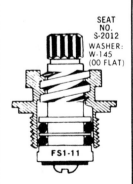

FS1-11C   FS1-11H
Fits SAYCO-NOLAND
OEM 45-D-12 Stem, 57-D-13 Bonnet for 280 Fitting

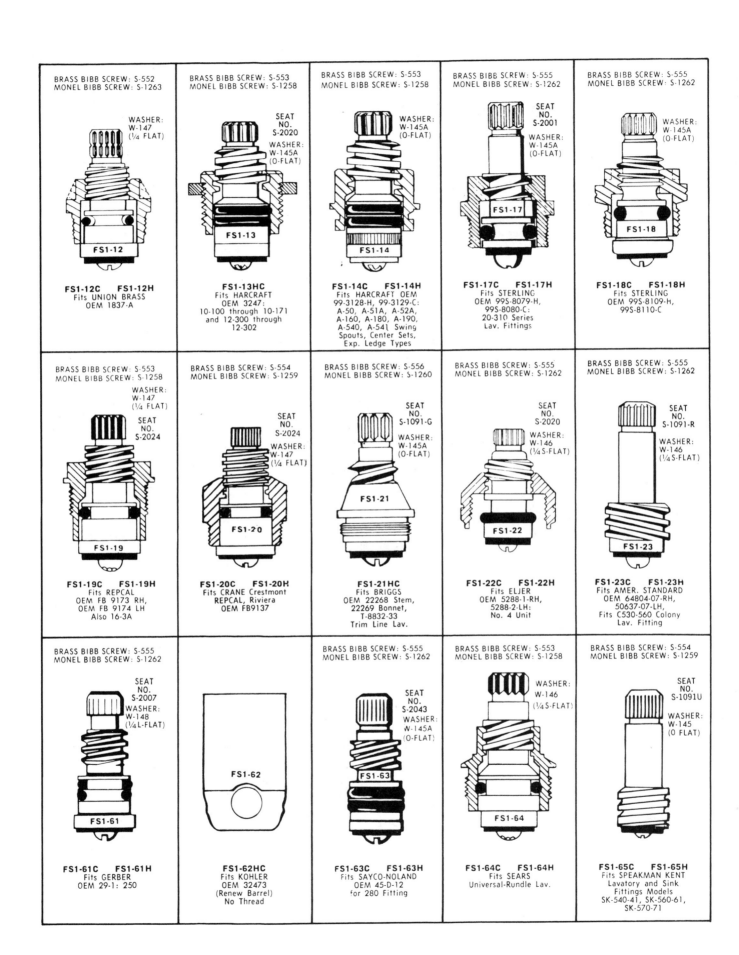

BRASS BIBB SCREW: S-552
MONEL BIBB SCREW: S-1263

WASHER:
W-147
(¼ FLAT)

FS1-12C   FS1-12H
Fits UNION BRASS
OEM 1837-A

BRASS BIBB SCREW: S-553
MONEL BIBB SCREW: S-1258

SEAT
NO.
S-2020
WASHER:
W-145A
(O-FLAT)

FS1-13HC
Fits HARCRAFT
OEM 3247:
10-100 through 10-171
and 12-300 through
12-302

BRASS BIBB SCREW: S-553
MONEL BIBB SCREW: S-1258

WASHER:
W-145A
(O-FLAT)

FS1-14C   FS1-14H
Fits HARCRAFT OEM
99-3128-H, 99-3129-C:
A-50, A-51A, A-52A,
A-160, A-180, A-190,
A-540, A-541 Swing
Spouts, Center Sets,
Exp. Ledge Types

BRASS BIBB SCREW: S-555
MONEL BIBB SCREW: S-1262

SEAT
NO.
S-2001
WASHER:
W-145A
(O-FLAT)

FS1-17C   FS1-17H
Fits STERLING
OEM 99S-8079-H,
99S-8080-C:
20-310 Series
Lav. Fittings

BRASS BIBB SCREW: S-555
MONEL BIBB SCREW: S-1262

WASHER:
W-145A
(O-FLAT)

FS1-18C   FS1-18H
Fits STERLING
OEM 99S-8109-H,
99S-8110-C

BRASS BIBB SCREW: S-553
MONEL BIBB SCREW: S-1258

WASHER:
W-147
(¼ FLAT)
SEAT
NO.
S-2024

FS1-19C   FS1-19H
Fits REPCAL
OEM FB 9173 RH,
OEM FB 9174 LH
Also 16-3A

BRASS BIBB SCREW: S-554
MONEL BIBB SCREW: S-1259

SEAT
NO.
S-2024
WASHER:
W-147
(¼ FLAT)

FS1-20C   FS1-20H
Fits CRANE Crestmont
REPCAL, Riviera
OEM FB9137

BRASS BIBB SCREW: S-556
MONEL BIBB SCREW: S-1260

SEAT
NO.
S-1091-G
WASHER:
W-145A
(O-FLAT)

FS1-21HC
Fits BRIGGS
OEM 22268 Stem,
22269 Bonnet,
T-8832-33
Trim Line Lav.

BRASS BIBB SCREW: S-555
MONEL BIBB SCREW: S-1262

SEAT
NO.
S-2020
WASHER:
W-146
(¼S-FLAT)

FS1-22C   FS1-22H
Fits ELJER
OEM 5288-1-RH,
5288-2-LH:
No. 4 Unit

BRASS BIBB SCREW: S-555
MONEL BIBB SCREW: S-1262

SEAT
NO.
S-1091-R
WASHER:
W-146
(¼S-FLAT)

FS1-23C   FS1-23H
Fits AMER. STANDARD
OEM 64804-07-RH,
50637-07-LH,
Fits C530-560 Colony
Lav. Fitting

BRASS BIBB SCREW: S-555
MONEL BIBB SCREW: S-1262

SEAT
NO.
S-2007
WASHER:
W-148
(¼L-FLAT)

FS1-61C   FS1-61H
Fits GERBER
OEM 29-1: 250

FS1-62HC
Fits KOHLER
OEM 32473
(Renew Barrel)
No Thread

BRASS BIBB SCREW: S-555
MONEL BIBB SCREW: S-1262

SEAT
NO.
S-2043
WASHER:
W-145A
(O-FLAT)

FS1-63C   FS1-63H
Fits SAYCO-NOLAND
OEM 45-D-12
for 280 Fitting

BRASS BIBB SCREW: S-553
MONEL BIBB SCREW: S-1258

WASHER:
W-146
(¼S-FLAT)

FS1-64C   FS1-64H
Fits SEARS
Universal-Rundle Lav.

BRASS BIBB SCREW: S-554
MONEL BIBB SCREW: S-1259

SEAT
NO.
S-1091U
WASHER:
W-145
(O FLAT)

FS1-65C   FS1-65H
Fits SPEAKMAN KENT
Lavatory and Sink
Fittings Models
SK-540-41, SK-560-61,
SK-570-71

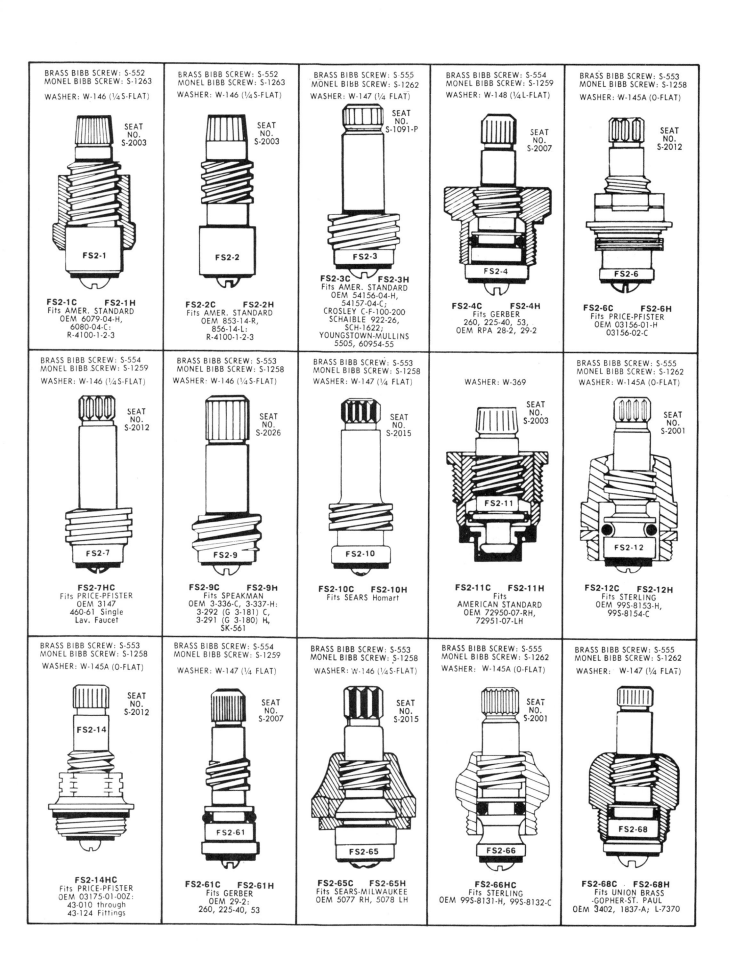

BRASS BIBB SCREW: S-552
MONEL BIBB SCREW: S-1263

WASHER: W-146 (¼S-FLAT)

SEAT NO. S-2003

FS2-1

FS2-1C   FS2-1H
Fits AMER. STANDARD
OEM 6079-04-H,
6080-04-C:
R-4100-1-2-3

BRASS BIBB SCREW: S-552
MONEL BIBB SCREW: S-1263

WASHER: W-146 (¼S-FLAT)

SEAT NO. S-2003

FS2-2

FS2-2C   FS2-2H
Fits AMER. STANDARD
OEM 853-14-R,
856-14-L:
R-4100-1-2-3

BRASS BIBB SCREW: S-555
MONEL BIBB SCREW: S-1262

WASHER: W-147 (¼ FLAT)

SEAT NO. S-1091-P

FS2-3

FS2-3C   FS2-3H
Fits AMER. STANDARD
OEM 54156-04-H,
54157-04-C;
CROSLEY C-F-100-200
SCHAIBLE 922-26,
SCH-1622;
YOUNGSTOWN-MULLINS
5505, 60954-55

BRASS BIBB SCREW: S-554
MONEL BIBB SCREW: S-1259

WASHER: W-148 (¼L-FLAT)

SEAT NO. S-2007

FS2-4

FS2-4C   FS2-4H
Fits GERBER
260, 225-40, 53,
OEM RPA 28-2, 29-2

BRASS BIBB SCREW: S-553
MONEL BIBB SCREW: S-1258

WASHER: W-145A (0-FLAT)

SEAT NO. S-2012

FS2-6

FS2-6C   FS2-6H
Fits PRICE-PFISTER
OEM 03156-01-H
03156-02-C

BRASS BIBB SCREW: S-554
MONEL BIBB SCREW: S-1259

WASHER: W-146 (¼S-FLAT)

SEAT NO. S-2012

FS2-7

FS2-7HC
Fits PRICE-PFISTER
OEM 3147
460-61 Single
Lav. Faucet

BRASS BIBB SCREW: S-553
MONEL BIBB SCREW: S-1258

WASHER: W-146 (¼S-FLAT)

SEAT NO. S-2026

FS2-9

FS2-9C   FS2-9H
Fits SPEAKMAN
OEM 3-336-C, 3-337-H:
3-292 (G 3-181) C,
3-291 (G 3-180) H,
SK-561

BRASS BIBB SCREW: S-553
MONEL BIBB SCREW: S-1258

WASHER: W-147 (¼ FLAT)

SEAT NO. S-2015

FS2-10

FS2-10C   FS2-10H
Fits SEARS Homart

WASHER: W-369

SEAT NO. S-2003

FS2-11

FS2-11C   FS2-11H
Fits
AMERICAN STANDARD
OEM 72950-07-RH,
72951-07-LH

BRASS BIBB SCREW: S-555
MONEL BIBB SCREW: S-1262

WASHER: W-145A (0-FLAT)

SEAT NO. S-2001

FS2-12

FS2-12C   FS2-12H
Fits STERLING
OEM 99S-8153-H,
99S-8154-C

BRASS BIBB SCREW: S-553
MONEL BIBB SCREW: S-1258

WASHER: W-145A (0-FLAT)

SEAT NO. S-2012

FS2-14

FS2-14HC
Fits PRICE-PFISTER
OEM 03175-01-00Z:
43-010 through
43-124 Fittings

BRASS BIBB SCREW: S-554
MONEL BIBB SCREW: S-1259

WASHER: W-147 (¼ FLAT)

SEAT NO. S-2007

FS2-61

FS2-61C   FS2-61H
Fits GERBER
OEM 29-2:
260, 225-40, 53

BRASS BIBB SCREW: S-553
MONEL BIBB SCREW: S-1258

WASHER: W-146 (¼S-FLAT)

SEAT NO. S-2015

FS2-65

FS2-65C   FS2-65H
Fits SEARS-MILWAUKEE
OEM 5077 RH, 5078 LH

BRASS BIBB SCREW: S-555
MONEL BIBB SCREW: S-1262

WASHER: W-145A (0-FLAT)

SEAT NO. S-2001

FS2-66

FS2-66HC
Fits STERLING
OEM 99S-8131-H, 99S-8132-C

BRASS BIBB SCREW: S-555
MONEL BIBB SCREW: S-1262

WASHER: W-147 (¼ FLAT)

FS2-68

FS2-68C   FS2-68H
Fits UNION BRASS
-GOPHER-ST. PAUL
OEM 3402, 1837-A; L-7370

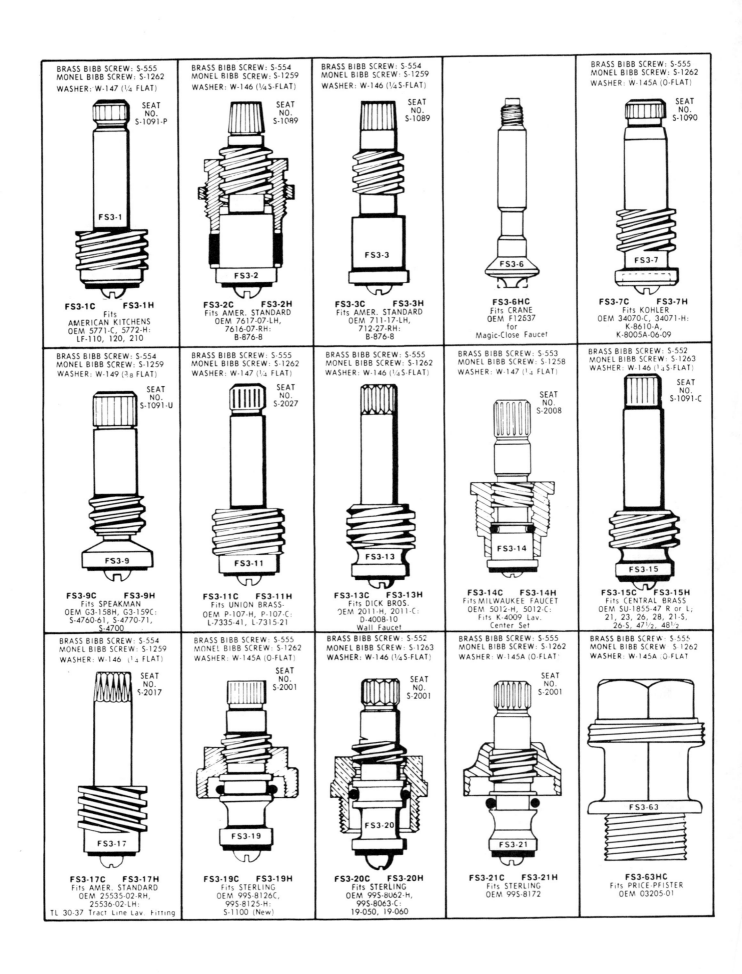

BRASS BIBB SCREW: S-555
MONEL BIBB SCREW: S-1262
WASHER: W-147 (¼ FLAT)

SEAT NO. S-1091-P

FS3-1

FS3-1C   FS3-1H
Fits
AMERICAN KITCHENS
OEM 5771-C, 5772-H:
LF-110, 120, 210

BRASS BIBB SCREW: S-554
MONEL BIBB SCREW: S-1259
WASHER: W-146 (¼S-FLAT)

SEAT NO. S-1089

FS3-2

FS3-2C   FS3-2H
Fits AMER. STANDARD
OEM 7617-07-LH,
7616-07-RH:
B-876-8

BRASS BIBB SCREW: S-554
MONEL BIBB SCREW: S-1259
WASHER: W-146 (¼S-FLAT)

SEAT NO. S-1089

FS3-3

FS3-3C   FS3-3H
Fits AMER. STANDARD
OEM 711-17-LH,
712-27-RH:
B-876-8

FS3-6

FS3-6HC
Fits CRANE
OEM F12637
for
Magic-Close Faucet

BRASS BIBB SCREW: S-555
MONEL BIBB SCREW: S-1262
WASHER: W-145A (O-FLAT)

SEAT NO. S-1090

FS3-7

FS3-7C   FS3-7H
Fits KOHLER
OEM 34070-C, 34071-H:
K-8610-A,
K-8005A-06-09

BRASS BIBB SCREW: S-554
MONEL BIBB SCREW: S-1259
WASHER: W-149 (⅜ FLAT)

SEAT NO. S-1091-U

FS3-9

FS3-9C   FS3-9H
Fits SPEAKMAN
OEM G3-158H, G3-159C:
S-4760-61, S-4770-71,
S-4700

BRASS BIBB SCREW: S-555
MONEL BIBB SCREW: S-1262
WASHER: W-147 (¼ FLAT)

SEAT NO. S-2027

FS3-11

FS3-11C   FS3-11H
Fits UNION BRASS-
OEM P-107-H, P-107-C:
L-7335-41, L-7315-21

BRASS BIBB SCREW: S-555
MONEL BIBB SCREW: S-1262
WASHER: W-146 (¼S-FLAT)

SEAT NO. S-2008

FS3-13

FS3-13C   FS3-13H
Fits DICK BROS.
OEM 2011-H, 2011-C:
D-4008-10
Wall Faucet

BRASS BIBB SCREW: S-553
MONEL BIBB SCREW: S-1258
WASHER: W-147 (¼ FLAT)

FS3-14

FS3-14C   FS3-14H
Fits MILWAUKEE FAUCET
OEM 5012-H, 5012-C:
Fits K-4009 Lav.
Center Set

BRASS BIBB SCREW: S-552
MONEL BIBB SCREW: S-1263
WASHER: W-146 (¼S-FLAT)

SEAT NO. S-1091-C

FS3-15

FS3-15C   FS3-15H
Fits CENTRAL BRASS
OEM SU-1855-47 R or L;
21, 23, 26, 28, 21-S,
26-S, 47½, 48½

BRASS BIBB SCREW: S-554
MONEL BIBB SCREW: S-1259
WASHER: W-146 (¼ FLAT)

SEAT NO. S-2017

FS3-17

FS3-17C   FS3-17H
Fits AMER. STANDARD
OEM 25535-02-RH,
25536-02-LH:
TL 30-37 Tract Line Lav. Fitting

BRASS BIBB SCREW: S-555
MONEL BIBB SCREW: S-1262
WASHER: W-145A (O-FLAT)

SEAT NO. S-2001

FS3-19

FS3-19C   FS3-19H
Fits STERLING
OEM 99S-8126C,
99S-8125-H:
S-1100 (New)

BRASS BIBB SCREW: S-552
MONEL BIBB SCREW: S-1263
WASHER: W-146 (¼S-FLAT)

SEAT NO. S-2001

FS3-20

FS3-20C   FS3-20H
Fits STERLING
OEM 99S-8062-H,
99S-8063-C:
19-050, 19-060

BRASS BIBB SCREW: S-555
MONEL BIBB SCREW: S-1262
WASHER: W-145A (O-FLAT)

SEAT NO. S-2001

FS3-21

FS3-21C   FS3-21H
Fits STERLING
OEM 99S-8172

BRASS BIBB SCREW: S-555
MONEL BIBB SCREW: S-1262
WASHER: W-145A (O-FLAT)

FS3-63

FS3-63HC
Fits PRICE-PFISTER
OEM 03205-01

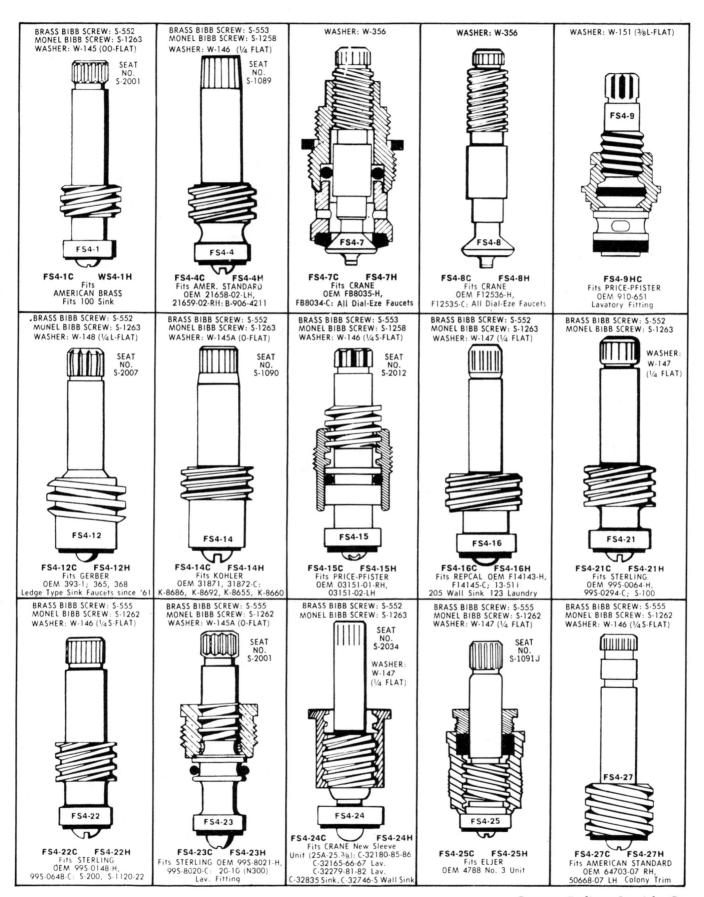

BRASS BIBB SCREW: S-552
MONEL BIBB SCREW: S-1263
WASHER: W-145 (00-FLAT)

SEAT NO. S-2001

FS4-1

FS4-1C   WS4-1H
Fits AMERICAN BRASS
Fits 100 Sink

BRASS BIBB SCREW: S-553
MONEL BIBB SCREW: S-1258
WASHER: W-146 (¼ FLAT)

SEAT NO. S-1089

FS4-4

FS4-4C   FS4-4H
Fits AMER. STANDARD
OEM 21658-02-LH,
21659-02-RH: B-906-4211

WASHER: W-356

FS4-7

FS4-7C   FS4-7H
Fits CRANE
OEM FB8035-H,
FB8034-C: All Dial-Eze Faucets

WASHER: W-356

FS4-8

FS4-8C   FS4-8H
Fits CRANE
OEM F12536-H,
F12535-C: All Dial-Eze Faucets

WASHER: W-151 (⅜L-FLAT)

FS4-9

FS4-9HC
Fits PRICE-PFISTER
OEM 910-651
Lavatory Fitting

BRASS BIBB SCREW: S-552
MONEL BIBB SCREW: S-1263
WASHER: W-148 (¼L-FLAT)

SEAT NO. S-2007

FS4-12

FS4-12C   FS4-12H
Fits GERBER
OEM 393-1; 365, 368
Ledge Type Sink Faucets since '61

BRASS BIBB SCREW: S-552
MONEL BIBB SCREW: S-1263
WASHER: W-145A (O-FLAT)

SEAT NO. S-1090

FS4-14

FS4-14C   FS4-14H
Fits KOHLER
OEM 31871, 31872-C:
K-8686, K-8692, K-8655, K-8660

BRASS BIBB SCREW: S-553
MONEL BIBB SCREW: S-1258
WASHER: W-146 (¼S-FLAT)

SEAT NO. S-2012

FS4-15

FS4-15C   FS4-15H
Fits PRICE-PFISTER
OEM 03151-01-RH,
03151-02-LH

BRASS BIBB SCREW: S-552
MONEL BIBB SCREW: S-1263
WASHER: W-147 (¼ FLAT)

FS4-16

FS4-16C   FS4-16H
Fits REPCAL OEM F14143-H,
F14145-C; 13-51i
205 Wall Sink 123 Laundry

BRASS BIBB SCREW: S-552
MONEL BIBB SCREW: S-1263

WASHER: W-147 (¼ FLAT)

FS4-21

FS4-21C   FS4-21H
Fits STERLING
OEM 99S-0064-H,
99S-0294-C; S-100

BRASS BIBB SCREW: S-555
MONEL BIBB SCREW: S-1262
WASHER: W-146 (¼S-FLAT)

FS4-22

FS4-22C   FS4-22H
Fits STERLING
OEM 99S-0148-H,
99S-0648-C; S-200, S-1120-22

BRASS BIBB SCREW: S-555
MONEL BIBB SCREW: S-1262
WASHER: W-145A (O-FLAT)

SEAT NO. S-2001

FS4-23

FS4-23C   FS4-23H
Fits STERLING OEM 99S-8021-H,
99S-8020-C: 20-10 (N300)
Lav. Fitting

BRASS BIBB SCREW: S-552
MONEL BIBB SCREW: S-1263

SEAT NO. S-2034

WASHER: W-147 (¼ FLAT)

FS4-24

FS4-24C   FS4-24H
Fits CRANE New Sleeve
Unit (25A-25.³⁄₈): C-32180-85-86
C-32165-66-67 Lav.
C-32279-81-82 Lav.
C-32835 Sink, C-32746-S Wall Sink

BRASS BIBB SCREW: S-555
MONEL BIBB SCREW: S-1262
WASHER: W-147 (¼ FLAT)

SEAT NO. S-1091J

FS4-25

FS4-25C   FS4-25H
Fits ELJER
OEM 4788 No. 3 Unit

BRASS BIBB SCREW: S-555
MONEL BIBB SCREW: S-1262
WASHER: W-146 (¼S-FLAT)

FS4-27

FS4-27C   FS4-27H
Fits AMERICAN STANDARD
OEM 64703-07 RH,
50668-07 LH Colony Trim

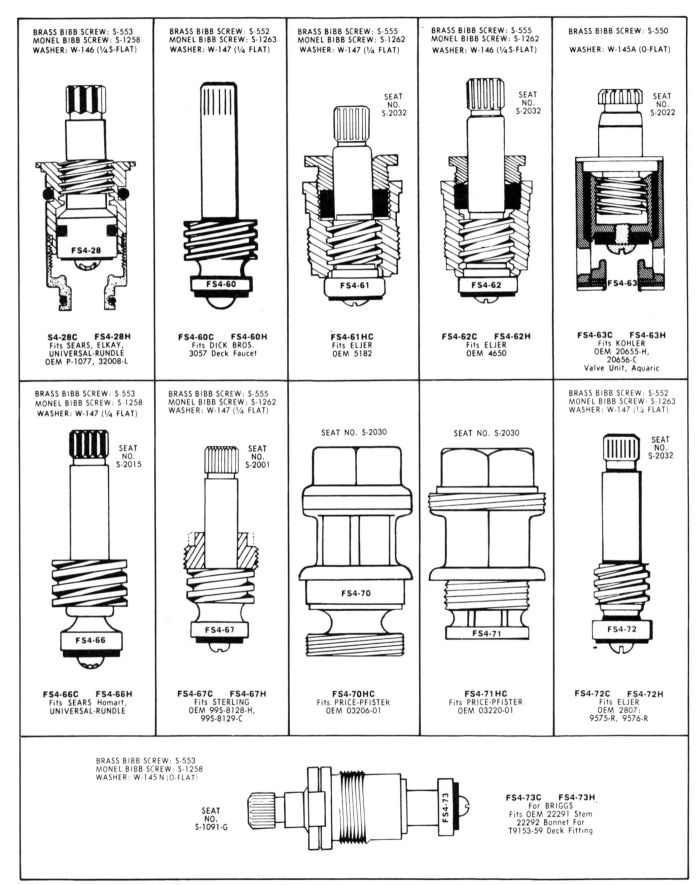

BRASS BIBB SCREW: S-553
MONEL BIBB SCREW: S-1258
WASHER: W-146 (¼S-FLAT)

FS4-28

S4-28C   FS4-28H
Fits SEARS, ELKAY,
UNIVERSAL-RUNDLE
OEM P-1077, 32008-L

BRASS BIBB SCREW: S-552
MONEL BIBB SCREW: S-1263
WASHER: W-147 (¼ FLAT)

FS4-60

FS4-60C   FS4-60H
Fits DICK BROS.
3057 Deck Faucet

BRASS BIBB SCREW: S-555
MONEL BIBB SCREW: S-1262
WASHER: W-147 (¼ FLAT)

SEAT NO. S-2032

FS4-61

FS4-61HC
Fits ELJER
OEM 5182

BRASS BIBB SCREW: S-555
MONEL BIBB SCREW: S-1262
WASHER: W-146 (¼S-FLAT)

SEAT NO. S-2032

FS4-62

FS4-62C   FS4-62H
Fits ELJER
OEM 4650

BRASS BIBB SCREW: S-550
WASHER: W-145A (O-FLAT)

SEAT NO. S-2022

FS4-63

FS4-63C   FS4-63H
Fits KOHLER
OEM 20655-H,
20656-C
Valve Unit, Aquaric

BRASS BIBB SCREW: S-553
MONEL BIBB SCREW: S-1258
WASHER: W-147 (¼ FLAT)

SEAT NO. S-2015

FS4-66

FS4-66C   FS4-66H
Fits SEARS Homart,
UNIVERSAL-RUNDLE

BRASS BIBB SCREW: S-555
MONEL BIBB SCREW: S-1262
WASHER: W-147 (¼ FLAT)

SEAT NO. S-2001

FS4-67

FS4-67C   FS4-67H
Fits STERLING
OEM 99S-8128-H,
99S-8129-C

SEAT NO. S-2030

FS4-70

FS4-70HC
Fits PRICE-PFISTER
OEM 03206-01

SEAT NO. S-2030

FS4-71

FS4-71HC
Fits PRICE-PFISTER
OEM 03220-01

BRASS BIBB SCREW: S-552
MONEL BIBB SCREW: S-1263
WASHER: W-147 (¼ FLAT)

SEAT NO. S-2032

FS4-72

FS4-72C   FS4-72H
Fits ELJER
OEM 2807;
9575-R, 9576-R

BRASS BIBB SCREW: S-553
MONEL BIBB SCREW: S-1258
WASHER: W-145 N (O-FLAT)

SEAT NO. S-1091-G

FS4-73

FS4-73C   FS4-73H
For BRIGGS
Fits OEM 22291 Stem
22292 Bonnet For
T9153-59 Deck Fitting

Courtesy Radiator Specialty Co.

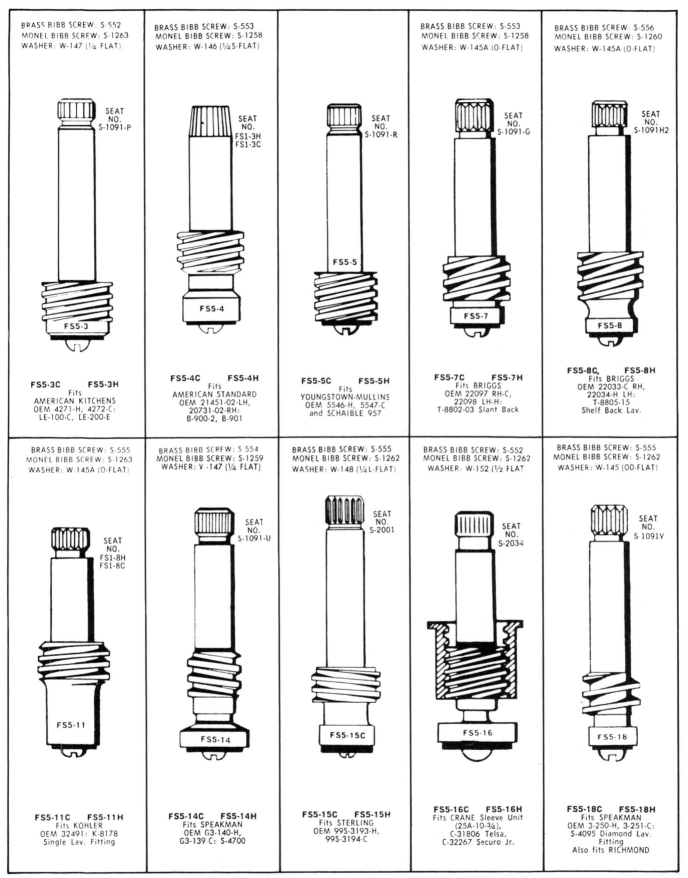

BRASS BIBB SCREW: S-552
MONEL BIBB SCREW: S-1263
WASHER: W-147 (¼ FLAT)

SEAT NO. S-1091-P

FS5-3

FS5-3C   FS5-3H
Fits
AMERICAN KITCHENS
OEM 4271-H, 4272-C:
LE-100-C, LE-200-E

BRASS BIBB SCREW: S-553
MONEL BIBB SCREW: S-1258
WASHER: W-146 (¼S-FLAT)

SEAT NO. FS1-3H FS1-3C

FS5-4

FS5-4C   FS5-4H
Fits
AMERICAN STANDARD
OEM 21451-02-LH,
20731-02-RH:
B-900-2, B-901

SEAT NO. S-1091-R

FS5-5

FS5-5C   FS5-5H
Fits
YOUNGSTOWN-MULLINS
OEM 5546-H, 5547-C
and SCHAIBLE 957

BRASS BIBB SCREW: S-553
MONEL BIBB SCREW: S-1258
WASHER: W-145A (O-FLAT)

SEAT NO. S-1091-G

FS5-7

FS5-7C   FS5-7H
Fits BRIGGS
OEM 22097 RH-C,
22098 LH-H:
T-8802-03 Slant Back

BRASS BIBB SCREW: S-556
MONEL BIBB SCREW: S-1260
WASHER: W-145A (O-FLAT)

SEAT NO. S-1091H2

FS5-8

FS5-8C   FS5-8H
Fits BRIGGS
OEM 22033-C RH,
22034-H LH:
T-8805-15
Shelf Back Lav.

BRASS BIBB SCREW: S-555
MONEL BIBB SCREW: S-1263
WASHER: W-145A (O-FLAT)

SEAT NO. FS1-8H FS1-8C

FS5-11

FS5-11C   FS5-11H
Fits KOHLER
OEM 32491: K-8178
Single Lav. Fitting

BRASS BIBB SCREW: S-554
MONEL BIBB SCREW: S-1259
WASHER: V-147 (¼ FLAT)

SEAT NO. S-1091-U

FS5-14

FS5-14C   FS5-14H
Fits SPEAKMAN
OEM G3-140-H,
G3-139-C: S-4700

BRASS BIBB SCREW: S-555
MONEL BIBB SCREW: S-1262
WASHER: W-148 (¼L-FLAT)

SEAT NO. S-2001

FS5-15C

FS5-15C   FS5-15H
Fits STERLING
OEM 99S-3193-H,
99S-3194-C

BRASS BIBB SCREW: S-552
MONEL BIBB SCREW: S-1262
WASHER: W-152 (½ FLAT)

SEAT NO. S-2034

FS5-16

FS5-16C   FS5-16H
Fits CRANE Sleeve Unit
(25A-10-¾),
C-31806 Telsa,
C-32267 Securo Jr.

BRASS BIBB SCREW: S-555
MONEL BIBB SCREW: S-1262
WASHER: W-145 (OO-FLAT)

SEAT NO. S-1091V

FS5-18

FS5-18C   FS5-18H
Fits SPEAKMAN
OEM 3-250-H, 3-251-C:
S-4095 Diamond Lav.
Fitting
Also fits RICHMOND

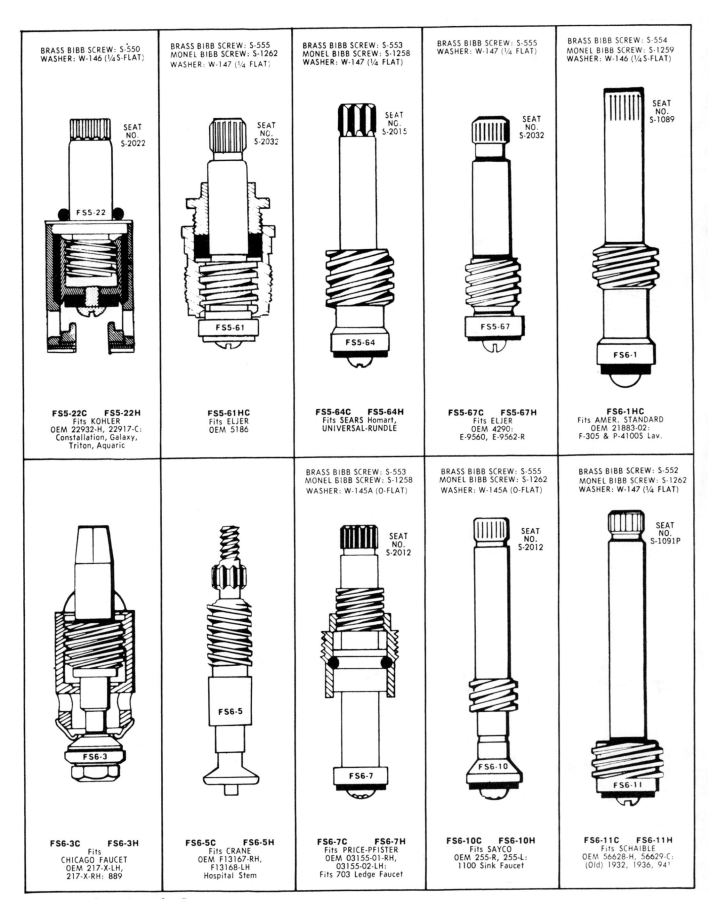

BRASS BIBB SCREW: S-550
WASHER: W-146 (1/4 S-FLAT)

SEAT NO. S-2022

FS5-22

FS5-22C   FS5-22H
Fits KOHLER
OEM 22932-H, 22917-C:
Constellation, Galaxy,
Triton, Aquaric

BRASS BIBB SCREW: S-555
MONEL BIBB SCREW: S-1262
WASHER: W-147 (1/4 FLAT)

SEAT NO. S-2032

FS5-61

FS5-61HC
Fits ELJER
OEM 5186

BRASS BIBB SCREW: S-553
MONEL BIBB SCREW: S-1258
WASHER: W-147 (1/4 FLAT)

SEAT NO. S-2015

FS5-64

FS5-64C   FS5-64H
Fits SEARS Homart,
UNIVERSAL-RUNDLE

BRASS BIBB SCREW: S-555
WASHER: W-147 (1/4 FLAT)

SEAT NO. S-2032

FS5-67

FS5-67C   FS5-67H
Fits ELJER
OEM 4290:
E-9560, E-9562-R

BRASS BIBB SCREW: S-554
MONEL BIBB SCREW: S-1259
WASHER: W-146 (1/4S-FLAT)

SEAT NO. S-1089

FS6-1

FS6-1HC
Fits AMER. STANDARD
OEM 21883-02:
F-305 & P-4100S Lav.

FS6-3

FS6-3C   FS6-3H
Fits
CHICAGO FAUCET
OEM 217-X-LH,
217-X-RH: 889

FS6-5

FS6-5C   FS6-5H
Fits CRANE
OEM F13167-RH,
F13168-LH
Hospital Stem

BRASS BIBB SCREW: S-553
MONEL BIBB SCREW: S-1258
WASHER: W-145A (O-FLAT)

SEAT NO. S-2012

FS6-7

FS6-7C   FS6-7H
Fits PRICE-PFISTER
OEM 03155-01-RH,
03155-02-LH:
Fits 703 Ledge Faucet

BRASS BIBB SCREW: S-555
MONEL BIBB SCREW: S-1262
WASHER: W-145A (O-FLAT)

SEAT NO. S-2012

FS6-10

FS6-10C   FS6-10H
Fits SAYCO
OEM 255-R, 255-L:
1100 Sink Faucet

BRASS BIBB SCREW: S-552
MONEL BIBB SCREW: S-1262
WASHER: W-147 (1/4 FLAT)

SEAT NO. S-1091P

FS6-11

FS6-11C   FS6-11H
Fits SCHAIBLE
OEM 56628-H, 56629-C:
(Old) 1932, 1936, 94¹

Courtesy Radiator Specialty Co.

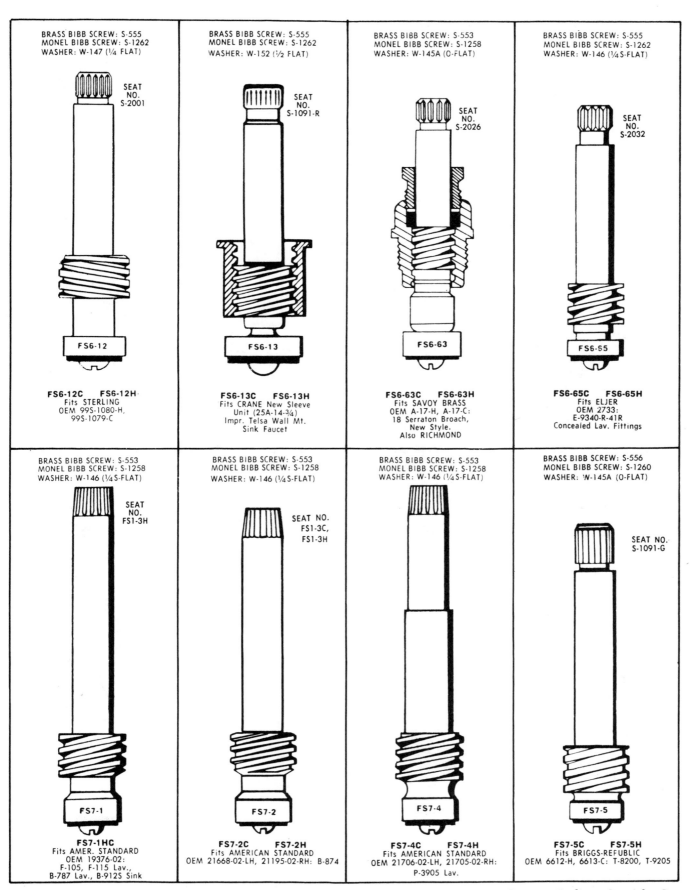

BRASS BIBB SCREW: S-555
MONEL BIBB SCREW: S-1262
WASHER: W-147 (¼ FLAT)

SEAT
NO.
S-2001

FS6-12

FS6-12C   FS6-12H·
Fits STERLING
OEM 99S-1080-H,
99S-1079-C

BRASS BIBB SCREW: S-555
MONEL BIBB SCREW: S-1262
WASHER: W-152 (½ FLAT)

SEAT
NO.
S-1091-R

FS6-13

FS6-13C   FS6-13H
Fits CRANE New Sleeve
Unit (25A-14-¾)
Impr. Telsa Wall Mt.
Sink Faucet

BRASS BIBB SCREW: S-553
MONEL BIBB SCREW: S-1258
WASHER: W-145A (O-FLAT)

SEAT
NO.
S-2026

FS6-63

FS6-63C   FS6-63H
Fits SAVOY BRASS
OEM A-17-H, A-17-C:
18 Serraton Broach,
New Style.
Also RICHMOND

BRASS BIBB SCREW: S-555
MONEL BIBB SCREW: S-1262
WASHER: W-146 (¼ S-FLAT)

SEAT
NO.
S-2032

FS6-65

FS6-65C   FS6-65H
Fits ELJER
OEM 2733:
E-9340-R-41R
Concealed Lav. Fittings

BRASS BIBB SCREW: S-553
MONEL BIBB SCREW: S-1258
WASHER: W-146 (¼S-FLAT)

SEAT
NO.
FS1-3H

FS7-1

FS7-1HC
Fits AMER. STANDARD
OEM 19376-02:
F-105, F-115 Lav.,
B-787 Lav., B-912S Sink

BRASS BIBB SCREW: S-553
MONEL BIBB SCREW: S-1258
WASHER: W-146 (¼S-FLAT)

SEAT NO.
FS1-3C,
FS1-3H

FS7-2

FS7-2C   FS7-2H
Fits AMERICAN STANDARD
OEM 21668-02-LH, 21195-02-RH: B-874

BRASS BIBB SCREW: S-553
MONEL BIBB SCREW: S-1258
WASHER: W-146 (¼S-FLAT)

FS7-4

FS7-4C   FS7-4H
Fits AMERICAN STANDARD
OEM 21706-02-LH, 21705-02-RH:
P-3905 Lav.

BRASS BIBB SCREW: S-556
MONEL BIBB SCREW: S-1260
WASHER: W-145A (O-FLAT)

SEAT NO.
S-1091-G

FS7-5

FS7-5C   FS7-5H
Fits BRIGGS-REFUBLIC
OEM 6612-H, 6613-C: T-8200, T-9205

Courtesy Radiator Specialty Co.

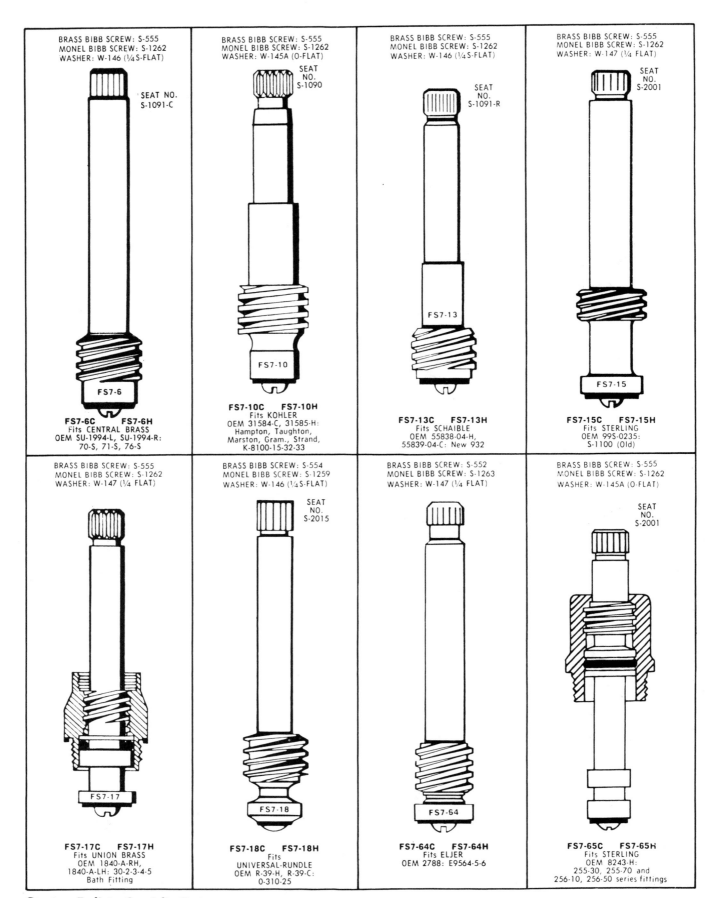

BRASS BIBB SCREW: S-555
MONEL BIBB SCREW: S-1262
WASHER: W-146 (¼ S-FLAT)

SEAT NO.
S-1091-C

FS7-6

FS7-6C   FS7-6H
Fits CENTRAL BRASS
OEM SU-1994-L, SU-1994-R:
70-S, 71-S, 76-S

BRASS BIBB SCREW: S-555
MONEL BIBB SCREW: S-1262
WASHER: W-145A (O-FLAT)

SEAT
NO.
S-1090

FS7-10

FS7-10C   FS7-10H
Fits KOHLER
OEM 31584-C, 31585-H:
Hampton, Taughton,
Marston, Gram., Strand,
K-8100-15-32-33

BRASS BIBB SCREW: S-555
MONEL BIBB SCREW: S-1262
WASHER: W-146 (¼ S-FLAT)

SEAT
NO.
S-1091-R

FS7-13

FS7-13C   FS7-13H
Fits SCHAIBLE
OEM 55838-04-H,
55839-04-C: New 932

BRASS BIBB SCREW: S-555
MONEL BIBB SCREW: S-1262
WASHER: W-147 (¼ FLAT)

SEAT
NO.
S-2001

FS7-15

FS7-15C   FS7-15H
Fits STERLING
OEM 99S-0235:
S-1100 (Old)

BRASS BIBB SCREW: S-555
MONEL BIBB SCREW: S-1262
WASHER: W-147 (¼ FLAT)

FS7-17

FS7-17C   FS7-17H
Fits UNION BRASS
OEM 1840-A-RH,
1840-A-LH: 30-2-3-4-5
Bath Fitting

BRASS BIBB SCREW: S-554
MONEL BIBB SCREW: S-1259
WASHER: W-146 (¼ S-FLAT)

SEAT
NO.
S-2015

FS7-18

FS7-18C   FS7-18H
Fits
UNIVERSAL-RUNDLE
OEM R-39-H, R-39-C:
0-310-25

BRASS BIBB SCREW: S-552
MONEL BIBB SCREW: S-1263
WASHER: W-147 (¼ FLAT)

FS7-64

FS7-64C   FS7-64H
Fits ELJER
OEM 2788: E9564-5-6

BRASS BIBB SCREW: S-555
MONEL BIBB SCREW: S-1262
WASHER: W-145A (O-FLAT)

SEAT
NO.
S-2001

FS7-65C   FS7-65H
Fits STERLING
OEM 8243-H:
255-30, 255-70 and
256-10, 256-50 series fittings

Courtesy Radiator Specialty Co.

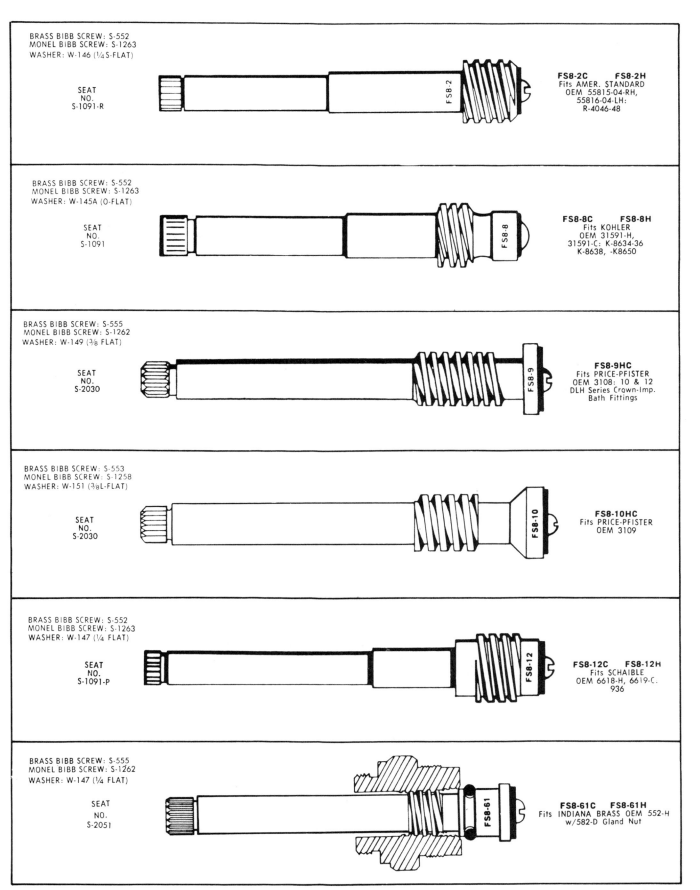

BRASS BIBB SCREW: S-552
MONEL BIBB SCREW: S-1263
WASHER: W-146 (¼S-FLAT)

SEAT
NO.
S-1091-R

FS8-2

FS8-2C   FS8-2H
Fits AMER. STANDARD
OEM 55815-04-RH,
55816-04-LH;
R-4046-48

BRASS BIBB SCREW: S-552
MONEL BIBB SCREW: S-1263
WASHER: W-145A (O-FLAT)

SEAT
NO.
S-1091

FS8-8

FS8-8C   FS8-8H
Fits KOHLER
OEM 31591-H,
31591-C; K-8634-36
K-8638, -K8650

BRASS BIBB SCREW: S-555
MONEL BIBB SCREW: S-1262
WASHER: W-149 (⅜ FLAT)

SEAT
NO.
S-2030

FS8-9

FS8-9HC
Fits PRICE-PFISTER
OEM 3108: 10 & 12
DLH Series Crown-Imp.
Bath Fittings

BRASS BIBB SCREW: S-553
MONEL BIBB SCREW: S-1258
WASHER: W-151 (⅜L-FLAT)

SEAT
NO.
S-2030

FS8-10

FS8-10HC
Fits PRICE-PFISTER
OEM 3109

BRASS BIBB SCREW: S-552
MONEL BIBB SCREW: S-1263
WASHER: W-147 (¼ FLAT)

SEAT
NO.
S-1091-P

FS8-12

FS8-12C   FS8-12H
Fits SCHAIBLE
OEM 6618-H, 6619-C.
936

BRASS BIBB SCREW: S-555
MONEL BIBB SCREW: S-1262
WASHER: W-147 (¼ FLAT)

SEAT
NO.
S-2051

FS8-61

FS8-61C   FS8-61H
Fits INDIANA BRASS OEM 552-H
w/582-D Gland Nut

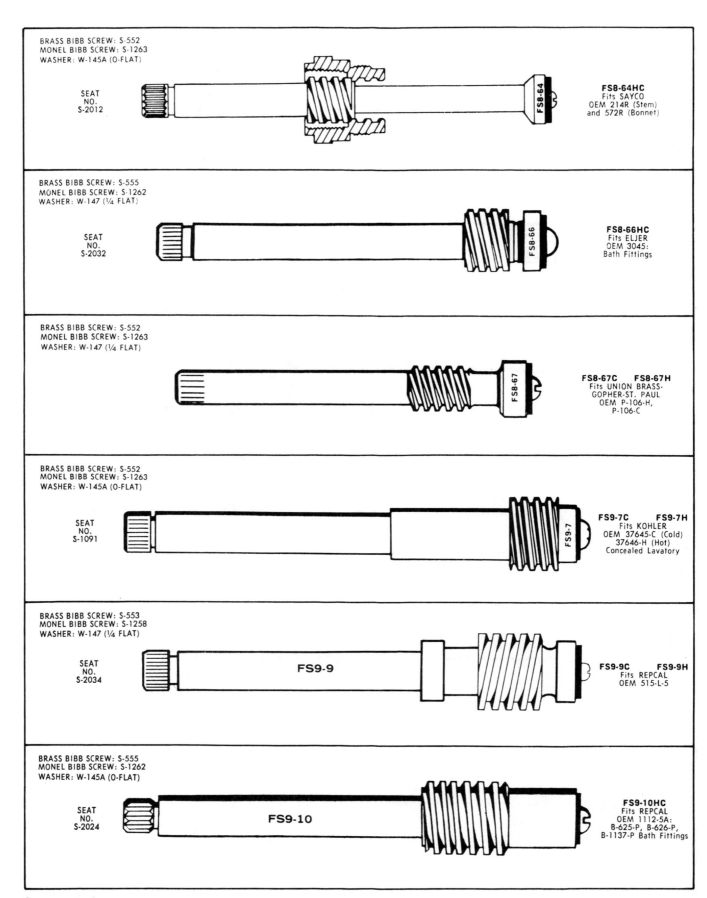

BRASS BIBB SCREW: S-552
MONEL BIBB SCREW: S-1263
WASHER: W-145A (O-FLAT)

SEAT NO. S-2012

FS8-64HC
Fits SAYCO
OEM 214R (Stem)
and 572R (Bonnet)

FS8-64

BRASS BIBB SCREW: S-555
MONEL BIBB SCREW: S-1262
WASHER: W-147 (¼ FLAT)

SEAT NO. S-2032

FS8-66HC
Fits ELJER
OEM 3045:
Bath Fittings

FS8-66

BRASS BIBB SCREW: S-552
MONEL BIBB SCREW: S-1263
WASHER: W-147 (¼ FLAT)

FS8-67C   FS8-67H
Fits UNION BRASS-
GOPHER-ST. PAUL
OEM P-106-H,
P-106-C

FS8-67

BRASS BIBB SCREW: S-552
MONEL BIBB SCREW: S-1263
WASHER: W-145A (O-FLAT)

SEAT NO. S-1091

FS9-7C   FS9-7H
Fits KOHLER
OEM 37645-C (Cold)
37646-H (Hot)
Concealed Lavatory

FS9-7

BRASS BIBB SCREW: S-553
MONEL BIBB SCREW: S-1258
WASHER: W-147 (¼ FLAT)

SEAT NO. S-2034

FS9-9

FS9-9C   FS9-9H
Fits REPCAL
OEM 515-L-5

BRASS BIBB SCREW: S-555
MONEL BIBB SCREW: S-1262
WASHER: W-145A (O-FLAT)

SEAT NO. S-2024

FS9-10

FS9-10HC
Fits REPCAL
OEM 1112-5A:
B-625-P, B-626-P,
B-1137-P Bath Fittings

Courtesy Radiator Specialty Co.

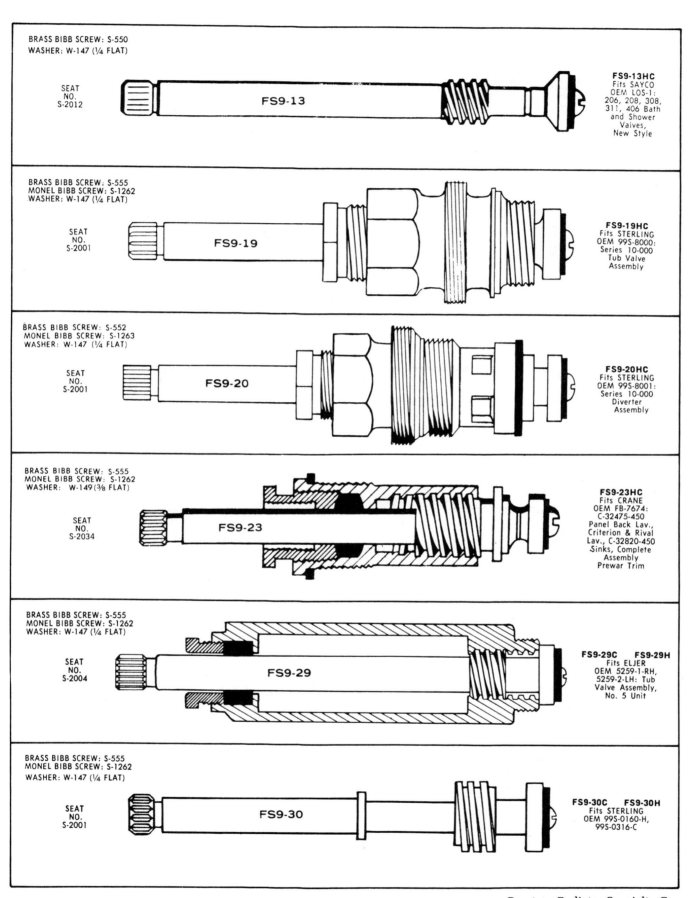

BRASS BIBB SCREW: S-550
WASHER: W-147 (¼ FLAT)

SEAT NO. S-2012

FS9-13

FS9-13HC
Fits SAYCO
OEM LOS-1:
206, 208, 308,
311, 406 Bath
and Shower
Valves,
New Style

BRASS BIBB SCREW: S-555
MONEL BIBB SCREW: S-1262
WASHER: W-147 (¼ FLAT)

SEAT NO. S-2001

FS9-19

FS9-19HC
Fits STERLING
OEM 99S-8000:
Series 10-000
Tub Valve
Assembly

BRASS BIBB SCREW: S-552
MONEL BIBB SCREW: S-1263
WASHER: W-147 (¼ FLAT)

SEAT NO. S-2001

FS9-20

FS9-20HC
Fits STERLING
OEM 99S-8001:
Series 10-000
Diverter
Assembly

BRASS BIBB SCREW: S-555
MONEL BIBB SCREW: S-1262
WASHER: W-149 (⅜ FLAT)

SEAT NO. S-2034

FS9-23

FS9-23HC
Fits CRANE
OEM FB-7674:
C-32475-450
Panel Back Lav.,
Criterion & Rival
Lav., C-32820-450
Sinks, Complete
Assembly
Prewar Trim

BRASS BIBB SCREW: S-555
MONEL BIBB SCREW: S-1262
WASHER: W-147 (¼ FLAT)

SEAT NO. S-2004

FS9-29

FS9-29C   FS9-29H
Fits ELJER
OEM 5259-1-RH,
5259-2-LH: Tub
Valve Assembly,
No. 5 Unit

BRASS BIBB SCREW: S-555
MONEL BIBB SCREW: S-1262
WASHER: W-147 (¼ FLAT)

SEAT NO. S-2001

FS9-30

FS9-30C   FS9-30H
Fits STERLING
OEM 99S-0160-H,
99S-0316-C

Courtesy Radiator Specialty Co.

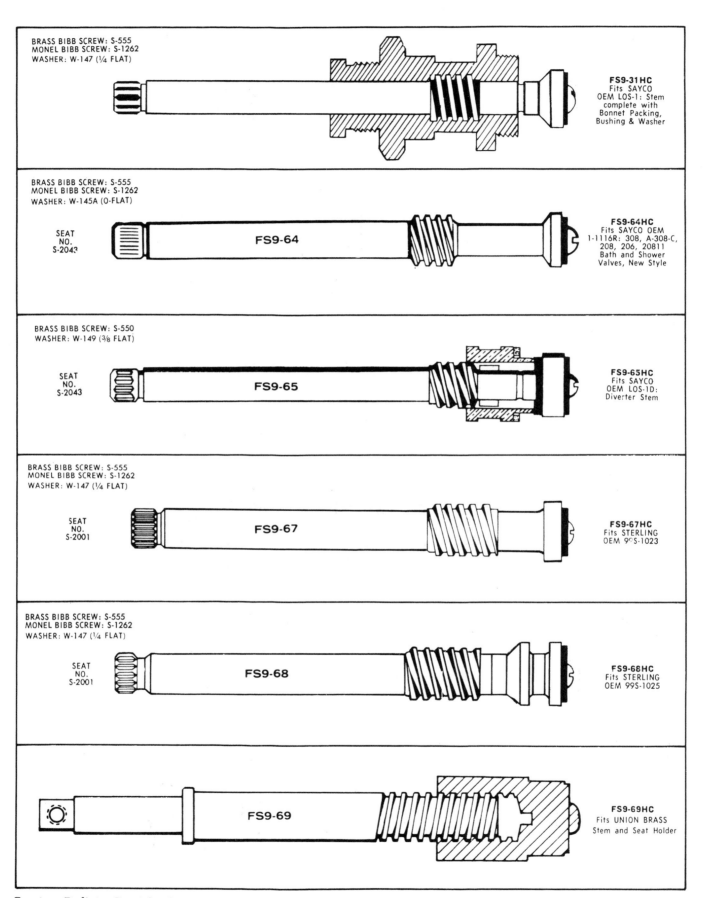

BRASS BIBB SCREW: S-555
MONEL BIBB SCREW: S-1262
WASHER: W-147 (¼ FLAT)

FS9-31HC
Fits SAYCO
OEM LOS-1: Stem
complete with
Bonnet Packing,
Bushing & Washer

BRASS BIBB SCREW: S-555
MONEL BIBB SCREW: S-1262
WASHER: W-145A (0-FLAT)

SEAT
NO.
S-2043

FS9-64

FS9-64HC
Fits SAYCO OEM
1-1116R: 308, A-308-C,
208, 206, 20811
Bath and Shower
Valves, New Style

BRASS BIBB SCREW: S-550
WASHER: W-149 (⅜ FLAT)

SEAT
NO.
S-2043

FS9-65

FS9-65HC
Fits SAYCO
OEM LOS-1D:
Diverter Stem

BRASS BIBB SCREW: S-555
MONEL BIBB SCREW: S-1262
WASHER: W-147 (¼ FLAT)

SEAT
NO.
S-2001

FS9-67

FS9-67HC
Fits STERLING
OEM 9°S-1023

BRASS BIBB SCREW: S-555
MONEL BIBB SCREW: S-1262
WASHER: W-147 (¼ FLAT)

SEAT
NO.
S-2001

FS9-68

FS9-68HC
Fits STERLING
OEM 99S-1025

FS9-69

FS9-69HC
Fits UNION BRASS
Stem and Seat Holder

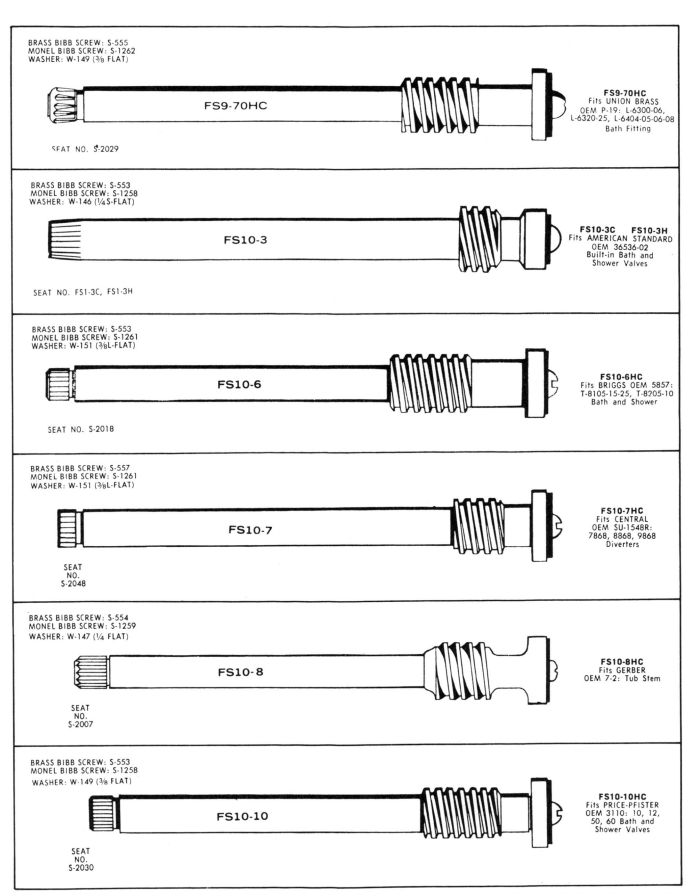

BRASS BIBB SCREW: S-555
MONEL BIBB SCREW: S-1262
WASHER: W-149 (3/8 FLAT)

FS9-70HC

SEAT NO. S-2029

FS9-70HC
Fits UNION BRASS
OEM P-19: L-6300-06,
L-6320-25, L-6404-05-06-08
Bath Fitting

BRASS BIBB SCREW: S-553
MONEL BIBB SCREW: S-1258
WASHER: W-146 (1/4 S-FLAT)

FS10-3

SEAT NO. FS1-3C, FS1-3H

FS10-3C   FS10-3H
Fits AMERICAN STANDARD
OEM 36536-02
Built-in Bath and
Shower Valves

BRASS BIBB SCREW: S-553
MONEL BIBB SCREW: S-1261
WASHER: W-151 (3/8 L-FLAT)

FS10-6

SEAT NO. S-2018

FS10-6HC
Fits BRIGGS OEM 5857:
T-8105-15-25, T-8205-10
Bath and Shower

BRASS BIBB SCREW: S-557
MONEL BIBB SCREW: S-1261
WASHER: W-151 (3/8 L-FLAT)

FS10-7

SEAT
NO.
S-2048

FS10-7HC
Fits CENTRAL
OEM SU-1548R:
7868, 8868, 9868
Diverters

BRASS BIBB SCREW: S-554
MONEL BIBB SCREW: S-1259
WASHER: W-147 (1/4 FLAT)

FS10-8

SEAT
NO.
S-2007

FS10-8HC
Fits GERBER
OEM 7-2: Tub Stem

BRASS BIBB SCREW: S-553
MONEL BIBB SCREW: S-1258
WASHER: W-149 (3/8 FLAT)

FS10-10

SEAT
NO.
S-2030

FS10-10HC
Fits PRICE-PFISTER
OEM 3110: 10, 12,
50, 60 Bath and
Shower Valves

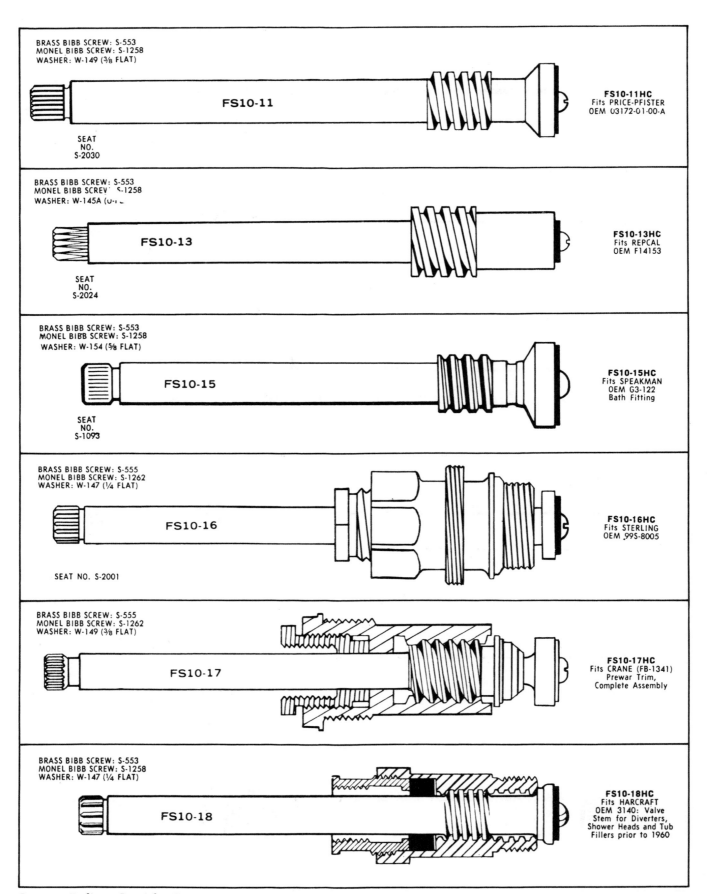

BRASS BIBB SCREW: S-553
MONEL BIBB SCREW: S-1258
WASHER: W-149 (3/8 FLAT)

FS10-11

FS10-11HC
Fits PRICE-PFISTER
OEM 03172-01-00-A

SEAT
NO.
S-2030

BRASS BIBB SCREW: S-553
MONEL BIBB SCREW: S-1258
WASHER: W-145A (O-R)

FS10-13

FS10-13HC
Fits REPCAL
OEM F14153

SEAT
NO.
S-2024

BRASS BIBB SCREW: S-553
MONEL BIBB SCREW: S-1258
WASHER: W-154 (5/8 FLAT)

FS10-15

FS10-15HC
Fits SPEAKMAN
OEM G3-122
Bath Fitting

SEAT
NO.
S-1093

BRASS BIBB SCREW: S-555
MONEL BIBB SCREW: S-1262
WASHER: W-147 (1/4 FLAT)

FS10-16

FS10-16HC
Fits STERLING
OEM 99S-8005

SEAT NO. S-2001

BRASS BIBB SCREW: S-555
MONEL BIBB SCREW: S-1262
WASHER: W-149 (3/8 FLAT)

FS10-17

FS10-17HC
Fits CRANE (FB-1341)
Prewar Trim,
Complete Assembly

BRASS BIBB SCREW: S-553
MONEL BIBB SCREW: S-1258
WASHER: W-147 (1/4 FLAT)

FS10-18

FS10-18HC
Fits HARCRAFT
OEM 3140: Valve
Stem for Diverters,
Shower Heads and Tub
Fillers prior to 1960

Courtesy Radiator Specialty Co.

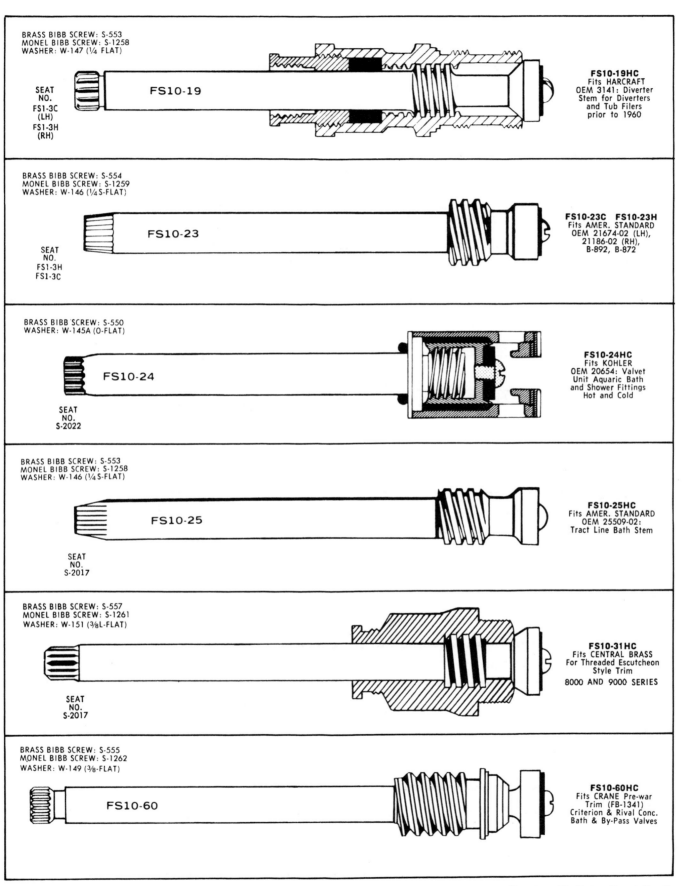

BRASS BIBB SCREW: S-553
MONEL BIBB SCREW: S-1258
WASHER: W-147 (¼ FLAT)

SEAT
NO.
FS1-3C
(LH)

FS1-3H
(RH)

FS10-19

FS10-19HC
Fits HARCRAFT
OEM 3141: Diverter
Stem for Diverters
and Tub Filers
prior to 1960

BRASS BIBB SCREW: S-554
MONEL BIBB SCREW: S-1259
WASHER: W-146 (¼ S-FLAT)

SEAT
NO.
FS1-3H
FS1-3C

FS10-23

FS10-23C FS10-23H
Fits AMER. STANDARD
OEM 21674-02 (LH),
21186-02 (RH),
B-892, B-872

BRASS BIBB SCREW: S-550
WASHER: W-145A (O-FLAT)

SEAT
NO.
S-2022

FS10-24

FS10-24HC
Fits KOHLER
OEM 20654: Valvet
Unit Aquaric Bath
and Shower Fittings
Hot and Cold

BRASS BIBB SCREW: S-553
MONEL BIBB SCREW: S-1258
WASHER: W-146 (¼ S-FLAT)

SEAT
NO.
S-2017

FS10-25

FS10-25HC
Fits AMER. STANDARD
OEM 25509-02:
Tract Line Bath Stem

BRASS BIBB SCREW: S-557
MONEL BIBB SCREW: S-1261
WASHER: W-151 (⅜ L-FLAT)

SEAT
NO.
S-2017

FS10-31

FS10-31HC
Fits CENTRAL BRASS
For Threaded Escutcheon
Style Trim
8000 AND 9000 SERIES

BRASS BIBB SCREW: S-555
MONEL BIBB SCREW: S-1262
WASHER: W-149 (⅜-FLAT)

FS10-60

FS10-60HC
Fits CRANE Pre-war
Trim (FB-1341)
Criterion & Rival Conc.
Bath & By-Pass Valves

Courtesy Radiator Specialty Co.

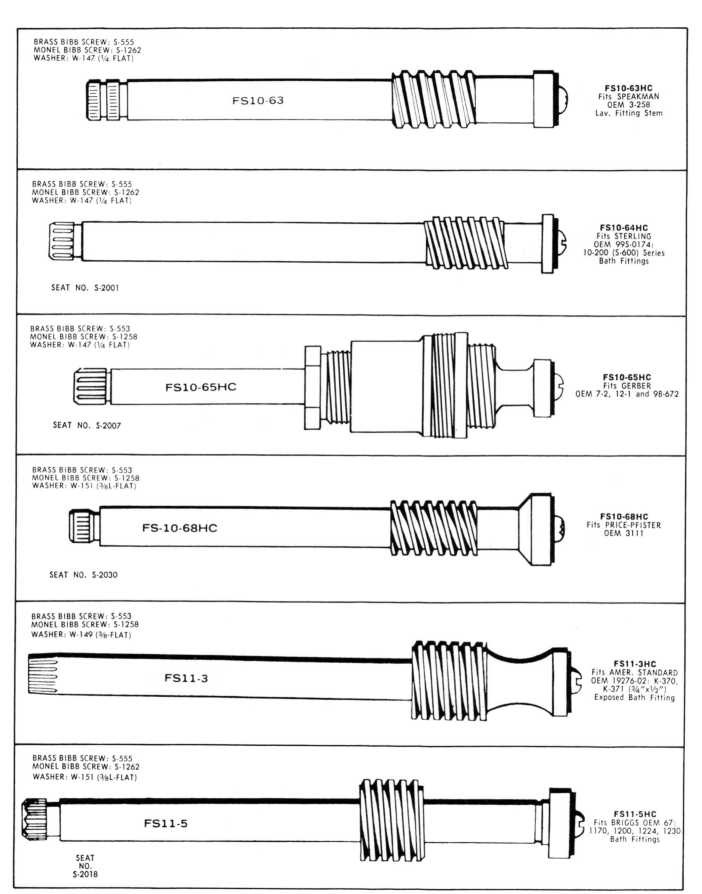

BRASS BIBB SCREW: S-555
MONEL BIBB SCREW: S-1262
WASHER: W-147 (¼ FLAT)

FS10-63

FS10-63HC
Fits SPEAKMAN
OEM 3-258
Lav. Fitting Stem

BRASS BIBB SCREW: S-555
MONEL BIBB SCREW: S-1262
WASHER: W-147 (¼ FLAT)

SEAT NO. S-2001

FS10-64HC
Fits STERLING
OEM 99S-0174:
10-200 (S-600) Series
Bath Fittings

BRASS BIBB SCREW: S-553
MONEL BIBB SCREW: S-1258
WASHER: W-147 (¼ FLAT)

FS10-65HC

SEAT NO. S-2007

FS10-65HC
Fits GERBER
OEM 7-2, 12-1 and 98-672

BRASS BIBB SCREW: S-553
MONEL BIBB SCREW: S-1258
WASHER: W-151 (⅜L-FLAT)

FS-10-68HC

SEAT NO. S-2030

FS10-68HC
Fits PRICE-PFISTER
OEM 3111

BRASS BIBB SCREW: S-553
MONEL BIBB SCREW: S-1258
WASHER: W-149 (⅜-FLAT)

FS11-3

FS11-3HC
Fits AMER. STANDARD
OEM 19276-02: K-370,
K-371 (¾"x½")
Exposed Bath Fitting

BRASS BIBB SCREW: S-555
MONEL BIBB SCREW: S-1262
WASHER: W-151 (⅜L-FLAT)

FS11-5

SEAT
NO.
S-2018

FS11-5HC
Fits BRIGGS OEM 67:
1170, 1200, 1224, 1230
Bath Fittings

Courtesy Radiator Specialty Co.

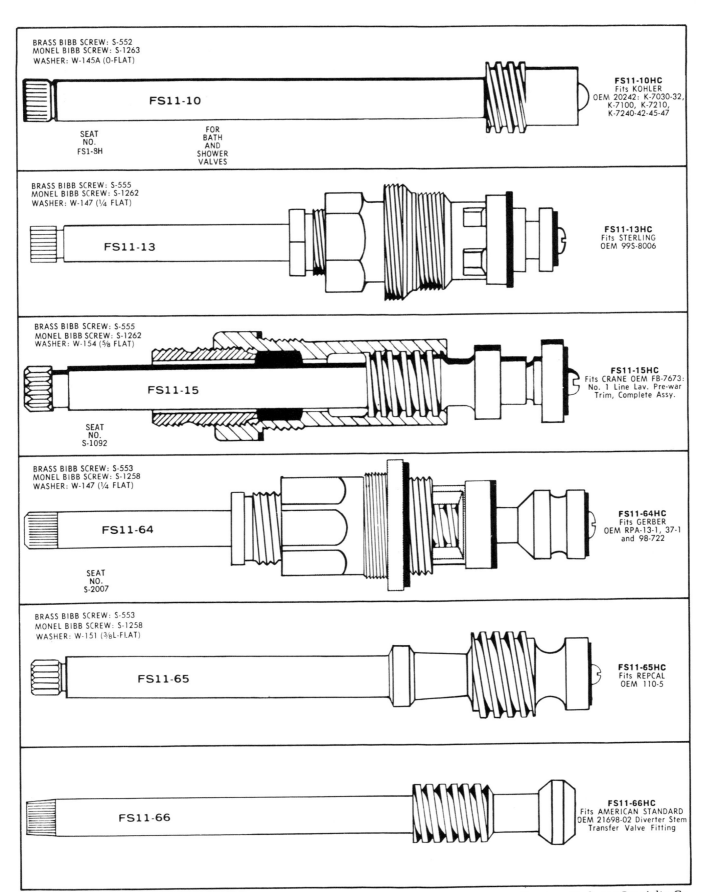

BRASS BIBB SCREW: S-552
MONEL BIBB SCREW: S-1263
WASHER: W-145A (O-FLAT)

FS11-10

SEAT
NO.
FS1-8H

FOR
BATH
AND
SHOWER
VALVES

FS11-10HC
Fits KOHLER
OEM 20242: K-7030-32,
K-7100, K-7210,
K-7240-42-45-47

BRASS BIBB SCREW: S-555
MONEL BIBB SCREW: S-1262
WASHER: W-147 (¼ FLAT)

FS11-13

FS11-13HC
Fits STERLING
OEM 99S-8006

BRASS BIBB SCREW: S-555
MONEL BIBB SCREW: S-1262
WASHER: W-154 (⅝ FLAT)

FS11-15

SEAT
NO.
S-1092

FS11-15HC
Fits CRANE OEM FB-7673:
No. 1 Line Lav. Pre-war
Trim, Complete Assy.

BRASS BIBB SCREW: S-553
MONEL BIBB SCREW: S-1258
WASHER: W-147 (¼ FLAT)

FS11-64

SEAT
NO.
S-2007

FS11-64HC
Fits GERBER
OEM RPA-13-1, 37-1
and 98-722

BRASS BIBB SCREW: S-553
MONEL BIBB SCREW: S-1258
WASHER: W-151 (⅜L-FLAT)

FS11-65

FS11-65HC
Fits REPCAL
OEM 110-5

FS11-66

FS11-66HC
Fits AMERICAN STANDARD
OEM 21698-02 Diverter Stem
Transfer Valve Fitting

Courtesy Radiator Specialty Co.

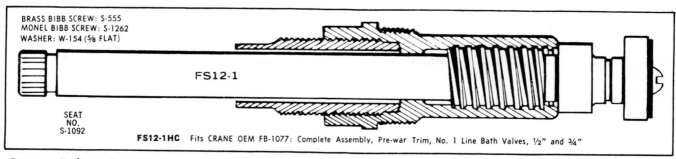

BRASS BIBB SCREW: S-555
MONEL BIBB SCREW: S-1262
WASHER: W-154 (⅝ FLAT)

FS12-1

SEAT
NO.
S-1092

FS12-1HC Fits CRANE OEM FB-1077: Complete Assembly, Pre-war Trim, No. 1 Line Bath Valves, ½" and ¾"

Courtesy Radiator Specialty Co.

"O" RINGS
PACKING RINGS FOR SWING SPOUT, SLIP-JOINTS, PUMP AND HYDRAULIC APPLICATIONS (ACTUAL SIZE ILLUSTRATIONS)

Part No.	Use	O.D.	I.D.	Wall
R-211	Fits Am. Std., Barnes, Central, Crane, Dick, Eljer, Indiana Brass, Milwaukee, Repcal, Moen, Price-Pfister, Schaible, Sears, Sterling	$\frac{3}{4}$	$\frac{9}{16}$	$\frac{3}{32}$
R-212	Fits Am. Std., Moen, Speakman, Union Brass	$\frac{13}{16}$	$\frac{5}{8}$	$\frac{3}{32}$
R-212A	Fits various makes	$\frac{13}{16}$	$\frac{11}{16}$	$\frac{1}{16}$
R-213	Fits Am. Kit., Barnes, Briggs, Gyro, Indiana Brass, Kohler, Moen, Price-Pfister, Repcal, Schaible, Sears, Tracy, Union Brass, Youngstown	$\frac{7}{8}$	$\frac{11}{16}$	$\frac{3}{32}$
R-214	Fits Crane	$\frac{15}{16}$	$\frac{3}{4}$	$\frac{3}{32}$
R-215	Fits Crane	1	$\frac{3}{4}$	$\frac{1}{8}$
R-216	Fits Eljer, Gyro, Moen	$1\frac{1}{16}$	$\frac{13}{16}$	$\frac{1}{8}$
R-217	Fits Am. Std. OEM 507-37, Moen 137, Repcal	$1\frac{1}{8}$	$\frac{7}{8}$	$\frac{1}{8}$
R-218	Fits Speakman	$1\frac{3}{16}$	$\frac{15}{16}$	$\frac{1}{8}$
R-219	Fits various makes	$1\frac{1}{4}$	1	$\frac{1}{8}$
R-220	Fits various makes	$1\frac{5}{16}$	$1\frac{1}{16}$	$\frac{1}{8}$
R-221	Fits Moen Model 52	$1\frac{3}{8}$	$1\frac{1}{8}$	$\frac{1}{8}$
R-222	Fits various makes	$1\frac{7}{16}$	$1\frac{3}{16}$	$\frac{1}{8}$
R-223	Fits Sears, Moen Model 42E	$1\frac{1}{2}$	$1\frac{1}{4}$	$\frac{1}{8}$
R-224	For special applications	$1\frac{9}{16}$	$1\frac{5}{16}$	$\frac{1}{4}$
R-225	For special applications	$1\frac{5}{8}$	$1\frac{3}{8}$	$\frac{1}{8}$
R-226	For special applications	$1\frac{11}{16}$	$1\frac{7}{16}$	$\frac{1}{8}$
R-227	For special applications	$1\frac{3}{4}$	$1\frac{1}{2}$	$\frac{1}{8}$
R-228	Fits Crane Magic-Close	$\frac{9}{16}$	$\frac{5}{16}$	$\frac{1}{8}$
R-990	Fits Am. Std. OEM 246-37 Aqua-Seal, Delta	$\frac{5}{8}$	$\frac{1}{2}$	$\frac{1}{16}$
R-991	Fits Delta, Tracy	$1\frac{1}{2}$	$1\frac{5}{16}$	$\frac{3}{32}$
R-992	Fits Am. Std. OEM 55234, Am. Kit., Crane, Crosley, Eljer, Kohler, Repcal, Schaible, Sears, Tracy, Yngstwn., Kohler OEM 29464	$1\frac{1}{16}$	$\frac{7}{8}$	$\frac{3}{32}$
R-993	Fits Central, Gerber, Moen, Indiana Brass, Univ.-Rundle	$\frac{11}{16}$	$\frac{1}{2}$	$\frac{3}{32}$
R-994	Fits Kohler OEM 34264	$1\frac{1}{16}$	$\frac{15}{16}$	$\frac{1}{16}$
R-995	Fits Am. Std., Kohler OEM 34300	$\frac{7}{8}$	$\frac{3}{4}$	$\frac{1}{16}$
R-996	Fits Alamark	$1\frac{3}{16}$	1	$\frac{3}{32}$
R-997	Fits Eljer OEM 5290	$\frac{9}{16}$	$\frac{5}{16}$	$\frac{1}{8}$
R-998	Fits Moen Model 42E	$1\frac{7}{16}$	$1\frac{5}{16}$	$\frac{1}{16}$
R-999	Fits Am. Std. OEM 886-17	$\frac{5}{8}$	$\frac{1}{2}$	$\frac{1}{16}$
R-1004	Fits Crane OEM F12376	$\frac{45}{64}$	$\frac{27}{64}$	$\frac{9}{64}$
R-1005	Fits Delta	$1\frac{1}{4}$	$1\frac{1}{8}$	$\frac{1}{16}$
R-1006	Fits Am. Std. OEM 12035, Schaible, Sears, Youngstown	$1\frac{3}{16}$	$1\frac{1}{16}$	$\frac{1}{16}$
R-1008	Fits Harcraft	$\frac{11}{16}$	$\frac{9}{16}$	$\frac{1}{16}$
R-1011	Fits Barnes	$\frac{23}{32}$	$\frac{15}{32}$	$\frac{9}{64}$
R-1015	Fits Eljer OEM 5285 Lusterline, Burlington 68-9	$1\frac{1}{4}$	$1\frac{1}{16}$	$\frac{3}{32}$

R-211 R-212 R-212A R-213 R-214 R-215
R-216 R-217 R-218 R-219
R-220 R-221 R-222 R-223
R-224 R-225 R-226 R-227 R-228
R-990 R-991 R-992 R-993 R-994 R-995
R-996 R-997 R-998 R-999
R-1004 R-1005 R-1006 R-1008 R-1011 R-1015

BIBB SCREWS

BRASS (SMALL HEADS)

Part No.	Description
S-550	(¼ x 8-32)
S-552	(½ x 8-32)
S-553	(⅜ x 10-24)
S-554	(½ x 10-24)
S-555	(⅜ x 8-32 #7 Head)
S-556	(⅜ x 10-28)
S-557	(½ x 10-28)
S-558	(⅜ x 10-32)
S-559	(½ x 10-32)
S-560	(⅜ x 6-32)
S-561	(½ x 6-32)

MONEL (SELF-LOCKING)

Will not corrode or crystalize. The head of MONEL screw is small so it won't touch seat of faucet. The nylon insert plug locks the screw in at proper depth. There is no chance for screw to come loose.

Part No.	Description
S-1256	(⅜x6-32) Gr.
S-1257	(½x6-32)
S-1258	(⅜x10-24)
S-1259	(½x10-24)
S-1260	(⅜x10-28)
S-1261	(½x10-28)
S-1262	(⅜x8-32)
S-1263	(½x8-32)
S-1264	(⅜x10-32)
S-1265	(½x10-32)

RENEW SEATS (ACTUAL SIZE ILLUSTRATIONS)

Part No.	Manufacturer of Faucet	O.E.M. Part No.	Specifications O.D. Thd. (Thd.) Ht.
FS1-3C	Amer. Standard	20563-08	Barrel Seat Cold Side (L.H. Thread)w O-Rings
FS1-3H	Amer. Standard	20336-08	Barrel Seat Hot Side (R.H. Thread)w O-Rings
FS1-8H	Kohler	39717	Barrel Seat Hot Side (L.H. Thread)
FS1-8C	Kohler	32462	Barrel Seat Cold Side (R.H. Thread)
FS1-8AC	Kohler	32462	Barrel Seat Cold Side (R.H. Thread)
FS1-8AH	Kohler	39717	Barrel Seat Hot Side (L.H. Thread)
FS1-62HC	Kohler	32473	Barrel Seat (Not Threaded)
S-1089	Amer. Standard	174-14	24 - 35/64 - 3/8
S-1090	Kohler	40602	27 - 1/2 - 25/64
S-1091	Kohler	33345	27 - 5/8 - 25/64
S-1091A	Dick Brothers	3055/4010	27 - 9/16 - 7/16
S-1091C	Central Brass	263-B	24 - 1/2 - 15/32
S-1091G	Briggs-Republic	22092	24 - 1/2 - 15/32
S-1091H	Briggs-Republic	8777	20 - 9/16 - 3/8
S-1091H2	Briggs-Republic	22040	20 - 1/2 - 3/4
S-1091J	Eljer	2734	27 - 1/2 - 1/2
S-1091P	Amer. Standard Amer. Kitchens Crosley Eljer Schaible Tracy Youngstown	56534-07 6534 6534 .... 6534 8662-8663 6534	20 - 1/2 - 3/8
S-1091R	Amer. Standard Schaible Youngstown	57281-07 5506 5506	20 - 1/2 - 7/16
S-1091S	Tracy	8435	20 - 5/8 - 7/16
S-1091S1	Amer. Standard Schaible Sears Youngstown	12002-07	20 - 35/64 - 5/16
S-1091U	Speakman	05-0305 Fmrly. S-5460	28 - 9/16 - 3/8

Illustrations labeled: S-1091H, S-1091H2, S-1091J, S-1091P, S-1091R, S-1091S, S-1091S1, S-1091U, FS1-3C, FS1-3H, FS1-62HC, S-1089, S-1090, FS1-8C, FS1-8H, S-1091, S-1091A, FS1-8AC, FS1-8AH, S-1091C, S-1091G, S-1091IJ

Courtesy Radiator Specialty Co.

RENEW SEATS (ACTUAL SIZE ILLUSTRATIONS)

Part No.	Manufacturer of Faucet	O.E.M. Part No.	Specifications Thd. - O.D.(Thd.) - Ht.		
S-1092	Crane	F-5914	22	43/64	27/64
S-1093	Burlington Br.	6-1	24	17/32	23/64
S-1094	Speakman	5-451	24	5/8	7/16
S-2001	Amer. Kitchen Sterling United Valve	802464 99S-0369	24	1/2	3/8
S-2001A	Empire Brass	20	24	1/2	11/32
S-2003	American Brass Amer. Standard	 862-14	24	7/16	13/32
S-2004	Barnes Eljer	7501 5257	20	1/2	11/32
S-2005	Barnes	7302	24	9/16	3/8
S-2006	Empire Brass	510	24	1/2	1¼
S-2007	Gerber	16	20	5/8	7/16
S-2008	Indiana Brass Milwaukee	638-0 964	27	1/2	3/8
S-2009	Milwaukee	3169	27	1/2	23/32
S-2011	Crane Savoy	 33	20	35/64	7/16
S-2011A	Savoy	524	20	1/2	13/32
S-2012	Price-Pfister Sayco	05716-01 6	20	1/2	3/8
S-2015	Univ.-Rundle	12R	24	17/32	3/8
S-2016	Wolverine	586S	24	1/2	3/8
S-2017	Amer. Standard	1848-07	18	17/32	3/8
S-2018	Briggs-Republic	91	20	5/8	27/32
S-2019	Gerber	98	20	5/8	1⅛
S-2020	Harcraft	6311	20	9/16	3/8
S-2021	Indiana	550-0	27	9/16	1⅝
S-2022	Kohler	23004	No Thd.	25/32	21/64
S-2023	Harcraft	5377	20	3/4	3/8
S-2024	Repcal	1112-18A	No Thd.	5/8	1/4
S-2025	Repcal	F14301 1112-18B	18	9/16	3/8
S-2026 (S-2010)	Speakman Richmond	05-0775 Fmrly. S-5465 K-12	27	7/16	11/32
S-2027 (S-2013)	Crane Union Brass	 2601	18	1/2	5/16
S-2028	Union Brass	2606	18	1/2	1⁵/₃₂
S-2029	Union Brass	2602	18	5/8	15/32
S-2030	Price-Pfister	05713-01	18	21/32	3/8

S-1092
S-1093
S-1094
S-2001
S-2001A
S-2003
S-2004
S-2005
S-2006
S-2007
S-2008

S-2009
S-2011
S-2011A
S-2012
S-2015
S-2016
S-2017
S-2018
S-2019
S-2020

S-2021
S-2022
S-2023
S-2024
S-2025
S-2026
S-2027
S-2028
S-2029
S-2030

RENEW SEATS (ACTUAL SIZE ILLUSTRATIONS)

Part No.	Manufacturer of Faucet	O.E.M. Part No.	Thd.	O.D. (Thd.)	Ht.
S-2031	Eljer	2924	27	1/2	$1\frac{9}{32}$
S-2032	Eljer	2750	27	35/64	15/16
S-2033	Eljer	4703	27	1/2	$1\frac{19}{32}$
S-2034	Crane	F5913	22	9/16	13/32
S-2035	Crane	F5914	24	11/16	15/32
S-2036	Crane Speakman		24	11/16	11/32
S-2037	Crane Michigan Brass	F3364 13117	18	9/16	3/4
S-2038	Crane Eljer Price-Pfister Sayco	583 5257 5716-01 6A3	24	1/2	5/16
S-2039	Briggs-Republic	6762	24	9/16	7/16
S-2040	Briggs-Republic	22379	20	5/8	7/16
S-2041	Wolverine	587-S	24	19/32	25/64
S-2042	Kohler	22526	27	5/8	3/8
S-2043	Sayco	6-468-C-39	2G	1/2	3/4
S-2044	Repcal	1112-18	28	9/16	11/32
S-2045	Union Brass	2632	20	39/64	5/16
S-2046	Central	59	27	9/16	13/32
S-2047	Central	263	24	1/2	47/64
S-2048	Central	165	24	5/8	23/32
S-2049	Central	165/6500	24	39/64	27/64
S-2051	Indiana Brass	590-0	27	9/16	5/8
S-2052	For use with Chicago, Economy, Sexauer or Skinner Tappg. Tools		27	3/8	3/8
S-2053	For use with Chicago, Economy, Sexauer or Skinner Tappg. Tools		27	3/8	19/64
S-2054	For use with Chicago, Economy, Sexauer or Skinner Tappg. Tools		27	7/16	3/8
S-2055	For use with Chicago, Economy, Sexauer or Skinner Tappg. Tools		27	7/16	19/64
S-2056	For use with Chicago, Economy, Sexauer or Skinner Tappg. Tools		27	1/2	19/64
S-2057	For use with Chicago, Economy, Sexauer or Skinner Tappg. Tools		27	1/2	1/4
S-2058	For use with Chicago, Economy, Sexauer or Skinner Tappg. Tools		27	1/2	3/8
S-2059	For use with Chicago, Economy, Sexauer or Skinner Tappg. Tools		27	1/2	19/64
S-2060	For use with Chicago, Economy, Sexauer or Skinner Tappg. Tools		27	9/16	3/8
S-2061	For use with Chicago, Economy, Sexauer or Skinner Tappg. Tools		27	9/16	19/64
S-2062	For use with Chicago, Economy, Sexauer or Skinner Tappg. Tools		27	5/8	3/8

Pressure Fittings Solder-type for copper plumbing

600 COUPLING W/STOP
Copper x Copper

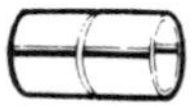

¼
⅜
⅜ x ¼
½
½ x ¼
½ x ⅜
¾
¾ x ⅝
¾ x ½
¾ x ⅜
1
1 x ¾
1 x ½
1¼

600-2 FITTING REDUCER
Slip Fit x Copper

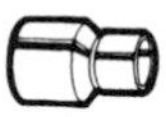

⅜ x ¼
½ x ⅜
¾ x ½
1 x ¾
1 x ½

601 REPAIR COUPLING LESS STOP
Copper x Copper

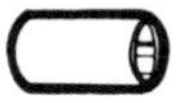

¼
⅜
½
¾
1

701-D DRAIN COUPLING
Copper x Copper

⅜
½
¾

701-2-D DRAIN COUPLING
Slip Fit x Copper

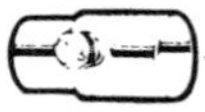

½
¾

717-D DRAIN CAP
Copper

½
¾

603 ADAPTER
Copper x Inside Thread

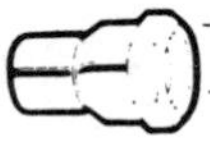

¼
⅜
⅜ x ½
½
½ x ¾
½ x ⅜
½ x ¼
⅝ x ¾
¾
¾ x 1
¾ x ½
1
1 x ¾

603-2 ADAPTER
Slip Fit x Inside Thread

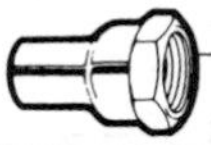

½

604 ADAPTER
Copper x Outside Thread

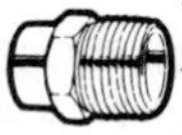

¼
⅜
⅜ x ¾
⅜ x ½
½
½ x 1
½ x ¾
½ x ⅜
¾
¾ x 1
¾ x ½
1
1 x ¾
1¼

604-2 ADAPTER
Slip Fit x Outside Thread

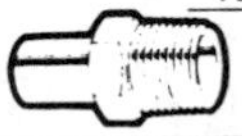

½

705 TEE — BASEBOARD
Copper x Inside Thread x Copper

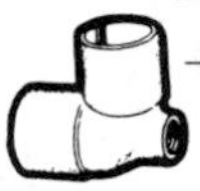

½ x ⅛ x ½
¾ x ⅛ x ¾

606 ELL — 45°
Copper x Copper

¼
⅜
½
¾
1
1¼
1½

606-2 ELL — 45°
Slip Fit x Copper

⅜
½
¾
1

607 ELL — 90°
Copper x Copper

¼
⅜
½
½ x ⅜
¾
¾ x ½
1
1 x ¾
1¼
1½

607-LT ELL — 90° LONG RADIUS
Copper x Copper

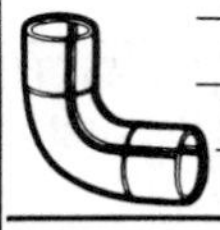

½
½ x ⅜
¾
1

607-2 ELL — 90°
Slip Fit x Copper

⅜
½
¾
1

607-3 ELL — 90°
Copper x Female

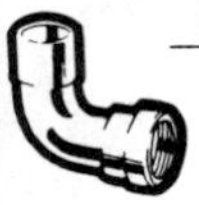

½

707-3 ELL — 90°
Copper x Inside Thread

½
½ x ¾
½ x ⅜
¾
¾ x ½

707-3-5 ELL — DROP EAR — 90°
Copper x Inside Thread

½
½ x ¾
½ x ⅜
¾

707-4 ELL — 90°
Copper x Outside Thread

½
½ x ¾
¾

707-4-5 ELL — DROP EAR — 90°
Copper x Outside Thread

½

707-5 ELL — DROP EAR — 90°

½
¾

708 90° SINK FITTINGS
Copper x Copper

½

611 TEE
Copper x Copper x Copper

¼
⅜
⅜ x ⅜ x ½
⅜ x ⅜ x ¼
½
½ x ½ x ¾
½ x ½ x ⅜
½ x ⅜ x ½
¾
¾ x ¾ x 1
¾ x ¾ x ½

¾ x ½ x ¾
¾ x ½ x ½
1
1 x 1 x ¾
1 x ¾ x 1
1 x ¾ x ¾
1 x ¾ x ½
1 x ½ x ¾

611-2 TEE — FITTING CONNECTION
Copper x Slip Fit x Copper

½
¾

712 TEE
Copper x Copper x Inside Thread

⅜
½
½ x ½ x ¾
½ x ½ x ⅜
¾
¾ x ¾ x ½

712-5 TEE — DROP EAR
Copper x Copper x Inside Thread

½
¾
¾ x ¾ x ½

714 TEE
Copper x Inside Thread x Copper

½

616 FITTING PLUG

⅜
½
¾

617 CAP-TUBE
Fits Over End of Tube

⅜
½
¾
1

Courtesy Nibcoware Div., Nibco, Inc.

Pressure Fittings Solder-type for copper plumbing

618 BUSHING
Slip Fit x Copper

3/8 x 1/4
1/2 x 3/8
3/4 x 1/2
1 x 3/4

619 AIR CHAMBER
Fits 1/2" Solder Cup

1/2 x 12
1/2 x 14*

*For Wisconsin Code

60 COPPER TUBING
Readi Cut™ Lengths

1/2 x 4'
1/2 x 6'
1/2 x 10'
3/4 x 4'
3/4 x 6'
3/4 x 10'

624 TUBE STRAP

3/8
1/2
3/4

724-5-A HY-SET HANGER COUPLING
Copper

1/2
3/4

733 UNION
Copper x Copper

3/8
1/2
3/4
1

736 CROSS-OVER
Copper x Copper

1/2
3/4

638 RETURN BEND
Copper x Copper

3/8
1/2
3/4

Flared Fittings
for copper piping

500 TUBE NUT

3/8
1/2
3/4
1

501 COUPLING
Copper x Copper

3/8
1/2
3/4
1

502 COUPLING — TWO PART W/RING
Copper x Copper

3/8
1/2
3/4

503 ADAPTER
Copper x Female

3/8
1/2
1/2 x 3/4
3/4
3/4 x 1/2
1

504 ADAPTER
Copper x Male

3/8
1/2
1/2 x 3/8
1/2 x 3/4
3/4
1

507 ELL — 90°
Copper x Copper

3/8
1/2
3/4
1

507-3 ELL — 90°
Copper x Female

3/8
1/2
3/4

507-4 ELL — 90°
Copper x Male

3/8
1/2
3/4

511 TEE
Copper x Copper x Copper

3/8
1/2
3/4
1

512 TEE
Copper x Copper x Female

1/2
3/4

516 CAP

3/8
1/2
3/4

523 FITTING REDUCER
Nut Seat x Copper

3/8 x 1/4
1/2 x 3/8
3/4 x 1/2
3/4 x 3/8

577-17 FLARED ANGLE STOP VALVE
Copper x Female

3/4
3/4 x 1/2

578-17 FLARED ANGLE STOP & WASTE VALVE
Copper x Female

3/4
3/4 x 1/2

DWV Fittings
Solder-type for copper drainage systems

901 COUPLING
Copper x Copper

1 1/4
1 1/2
1 1/2 x 1 1/4
2
2 x 1 1/2
2 x 1 1/4
3
3 x 1 1/2
3 x 1 1/4

901-RP COUPLING — REPAIR
Copper x Copper

1 1/4
1 1/2
2
3

901-2 BUSHING — EXTENDED
Fitting x Copper

1 1/2 x 1 1/4
3 x 1 1/2

801-2-T ADAPTER — TRAP
Fitting x O.D. Tube

1 1/2 x 1 1/2 OD
1 1/2 x 1 1/4 OD

901-7 ADAPTER — SLIP JOINT — TRAP
Copper x Slip Joint

1 1/4
1 1/2
1 1/2 x 1 1/4

DWV Fittings Solder-type for copper drainage systems

Fig. No. — Nominal Size

802 ADAPTER — TRAP
Copper x Outside Thread

1¼ x 1½
1½

903 ADAPTER
Copper x Inside Thread

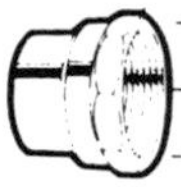

1¼
1½
2
3*

*Manufactured in Cast Bronze only. Specify Fig. No. 803

903-2 ADAPTER
Fitting x Inside Thread

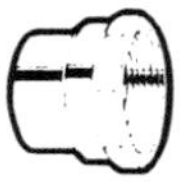

1¼
1½
2
3*

*Manufactured in Cast Bronze only. Specify Fig. No. 803-2

904 ADAPTER
Copper x Outside Thread

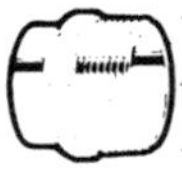

1¼
1½
2
3*

*Manufactured in Cast Bronze only. Specify Fig. No. 804

905 ADAPTER — SOIL PIPE
Copper x Spigot

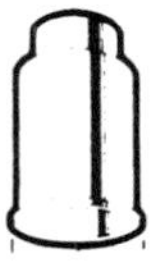

1½ x 2
2 x 2
2 x 3*
3 x 3*
3 x 4*

*Manufactured in Cast Bronze only. Specify Fig. No. 805

906 ELBOW — 45°
Copper x Copper

1¼
1½
2
3

Fig. No. — Nominal Size

906-2 ELBOW — 45°
Fitting x Copper

1¼
1½
2
3

907 ELBOW — 90°
Copper x Copper

1¼
1½
2
3

907-2 ELBOW — 90°
Fitting x Copper

1¼
1½
2
3

807-7 ELBOW — 90°
Copper x Slip Joint

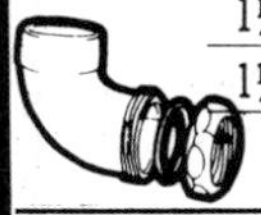

1½
1½ x 1¼

908 ELBOW — 22½°
Copper x Copper

1¼
1½
2
3

810-Y — 45°
Copper x Copper x Copper

1¼
1½
2
2 x 2 x 1½
2 x 1½ x 1½
3
3 x 3 x 1½
3 x 3 x 2
3 x 2 x 2

911 TEE
Copper x Copper x Copper

1¼
1½
2
2 x 2 x 1½
2 x 1½ x 2

Fig. No. — Nominal Size

2 x 1½ x 1½
3*
3 x 3 x 2*
3 x 3 x 1½*

*Manufactured in Cast Bronze only. Specify Fig. No. 811.

816-S CLEANOUT
Fitting x Cleanout W/Plug

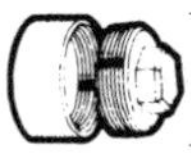

1¼
1½ x 1
2 x 1½
3 x 2½

818 A.S.A. PLUG
Threaded

1½
3

951 CLOSET FLANGE
Copper

3
4

851-C CLOSET FLANGE

4 x 3

960 ELBOW — 60°
Copper x Copper

1¼
1½
2
3

876 RETURN BEND W/CLEANOUT
Copper x Slip Joint

1½

878 RETURN BEND W/CLEANOUT
Copper x Copper x Cleanout

1½
2

Fig. No. — Nominal Size

880 P-TRAP W/CLEANOUT
Copper x Slip Joint x Cleanout W/Plug

1¼
1½
2

884 P-TRAP W/CLEANOUT
Copper x Copper x Cleanout W/Plug

1¼
1½
2
3

891 DRUM TRAP 3″ x 6″ SWIVEL DRUM
Copper x Copper

1½
2

892 P-TRAP W/UNION JOINT
Copper x Slip Joint

1½

892-3 P-TRAP W/UNION JOINT
Female x Slip Joint

1½

Accessories

Fig. No. — Nominal Size

518 FLARING TOOL

⅜
½
¾
1

755 FITTING CLEANING BRUSH

⅜
½
⅝
¾

Courtesy Nibcoware Div., Nibco, Inc.

Accessories

Fig. No.	Nominal Size
757 TUBE END CLEANING BRUSH	

3/8
1/2
3/4

765 SAND CLOTH
1½" Wide

10-Yard Roll
25-Yard Roll
50-Yard Roll

766 NOKORODE
Paste Flux

1.7 oz. can

1 lb. can

Valves Plumbing & Heating (for use with copper and iron pipe)

Fig. No.	Nominal Size

Gate Valves

T-22 U-VALVE
S-22
Female Thread x Female Thread
Copper x Copper

1/2
3/4
1
1¼
1½
2

T-22K U-VALVE W/CROSS HANDLE
Female Thread x Female Thread

1/2
3/4
1

T-29 FLAT TOP GATE VALVE
S-29
Female Thread x Female Thread
Copper x Copper

1/2
3/4
1
1¼
1½
2

Fig. No.	Nominal Size
T-180 RING GATE VALVE **S-180**	

Female Thread x Female Thread
Copper x Copper

3/8
1/2
3/4
1
1¼
1½
2

Ball Valves

T-580 RING BALL VALVE
S-580
Female Thread x Female Thread
Copper x Copper

1/2
3/4
1
1¼
1½
2

Check Valves

T-480 IN-LINE CHECK VALVE
S-480
Female Thread x Female Thread
Copper x Copper

1/2
3/4
1
1¼

Hex Shoulder Hose Bibbs

56 HOSE BIBB – S.S.S.
Male Thread x Hose

1/2
3/4

57 HOSE BIBB – S.O.T.
Male Thread x Hose

1/2
3/4

Fig. No.	Nominal Size

Angle Lawn Faucets

63 ANGLE SILL FAUCET W/FLANGE
Female Thread x Hose

1/2
3/4

763 ANGLE SILL FAUCET W/FLANGE
Copper x Hose

1/2
3/4

T-53 ANGLE SILL FAUCET
S-53
Female Thread x Hose
Copper x Hose

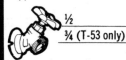

1/2
3/4 (T-53 only)

Garden Hose Valve

61 GARDEN HOSE VALVE
Female Thread x Hose

1/2
3/4

Frost Proof Lawn Faucets

52 FROST PROOF LAWN FAUCET
Threaded for Garden Hose

1/2 (½ Copper x ½ Male Thread)
3/4 (¾ Male Thread x ½ Female Thread)

8" length
10" length
12" length

Drain Valves
(Also for use with washing machine hook-ups.)

72 DRAIN VALVE
Copper x Hose

1/2
3/4

Fig. No.	Nominal Size
73 DRAIN VALVE	

Female Thread x Hose

1/2
3/4

74 DRAIN VALVE
Male Thread x Hose

1/2
3/4

4464 DRAIN VALVE
Compression x Hose

1/2

Non Kink Hose Faucets

54 NON KINK HOSE FAUCET
(Boiler or Drain Valve)
Male Thread x Hose

1/2
3/4

55 NON KINK HOSE FAUCET
(Boiler or Drain Valve)
Female Thread x Hose

1/2
3/4

Stop Valves
(Globe Pattern)

75 STOP VALVE
Female Thread x Female Thread

3/8
1/2
3/4
1

Valves Plumbing & Heating (for use with copper and iron pipe)

Fig. No.	Nominal Size
725 STOP VALVE	
Copper x Copper	
	3/8
	1/2
	3/4
	1

Stop & Waste Valves (Globe Pattern)

Fig. No.	Nominal Size
76 STOP & WASTE VALVE	
Female Thread x Female Thread	
	3/8
	1/2
	3/4
	1
726 STOP & WASTE VALVE	
Copper x Copper	
	3/8
	1/2
	3/4
	1
4476 STOP & WASTE VALVE	
Compression x Compression	
	1/2
	3/4

Gas Cocks

Fig. No.	Nominal Size
35 GAS COCK — FLAT HEAD	
Female Thread x Female Thread	
	3/8
	1/2
	3/4
	1
36 GAS COCK — LEVER HANDLE	
Female Thread x Female Thread	
	3/8
	1/2
	3/4
	1

Western States Valves

Fig. No.	Nominal Size
H9054 WASHING MACHINE VALVE FILLER — By-Pass Satin 1/2" Female Thread x 1/2" Male Thread w/3/4" Hose Outlet on bottom	1/2
H9055 WASHING MACHINE VALVE FILLER — Reversible By-Pass — Satin 1/2" Female Thread x 1/2" Male Thread w/3/4" Hose Outlet on side (Reversible either left or right.)	1/2
H9058 WASHING MACHINE VALVE FILLER Satin plated. Connect Automatic Washing Machine Hose to 1/2" Female Thread	1/2
H9059 WASHING MACHINE VALVE FILLER Satin 1/2" Female Thread water inlet w/3/4" hose outlet on bottom	1/2
H9073 EVAPORATIVE COOLER FAUCET 3/4" Female Hose Thread inlet x 3/4" Male Hose outlet. Tapped 1/8" Female Thread thru side wall	3/4

CPVC Hot & cold plastic fittings & pipe

Fig. No.	Nominal Size
4701 COUPLING	
Plastic x Plastic	
	1/2
	3/4
	3/4 x 1/2
4704 ADAPTER	
Plastic x Outside Thread	
	1/2
	3/4
4706 ELBOW — 45°	
Plastic x Plastic	
	1/2
	3/4
4707 ELBOW — 90°	
Plastic x Plastic	
	1/2
	3/4
	3/4 x 1/2
4707-2 ELBOW — 90°	
Fitting x Plastic	
	1/2
	3/4
4711 TEE	
Plastic x Plastic x Plastic	
	1/2
	3/4
	3/4 x 3/4 x 1/2
	3/4 x 1/2 x 3/4
	3/4 x 1/2 x 1/2
4717 CAP	
Plastic	
	1/2
	3/4
4718 BUSHING	
Fitting x Plastic	
	3/4 x 1/2
4724 STRAP — "SNAP-ON"	
(Polypropylene)	
	1/2
	3/4

Fig. No.	Nominal Size
4401 UNION COUPLING	
Compression x Compression	
	1/2
	1/2 x 1/4
	3/4
4403 ADAPTER	
Compression x Inside Thread	
	1/2
4404 ADAPTER	
Compression x Outside Thread	
	1/2
	3/4
4407-3-5 ELL — DROP EAR	
Compression x Inside Thread	
	1/2
4732-2 TRANSITION	
Fitting x Compression	
	1/2
	3/4
4733-4 UNION	
Plastic x Outside Thread	
	1/2
	1/2 x 3/4
	3/4
4464 WASHING MACHINE VALVE	
Compression x Hose	
	1/2
4476 STOP & WASTE VALVE	
Compression x Compression	
	1/2
	3/4
4776 STOP & WASTE VALVE	
CPVC to CPVC	
	1/2
	3/4

CPVC Hot & cold plastic fittings & pipe

Fig. No.	Nominal Size
47 CPVC PIPE	
	½
	¾

4798 SOLVENT
CPVC With Applicator

	¼ Pint
	½ Pint
	1 Pint

4899 PRIMER
PVC & CPVC With Applicator

| | ½ Pint |
| | 1 Pint |

DWV Fittings
for ABS and PVC drainage systems

Fig. No.	Nominal Size
4800 & 5800 ADAPTER — SOIL PIPE Plastic x Hub	
	2
	3
	4
5800 CLAY — ADAPTER — SOIL PIPE Plastic x Clay Pipe Hub	
	4
4800-SD & 5800-SD ADAPTER Plastic x ASTMD 2852 Plastic	
	3 x 4
4801 & 5801 COUPLING Plastic x Plastic	
	1¼
	1½
	1½ x 1¼
	2
	2 x 1½
	3
	3 x 1½

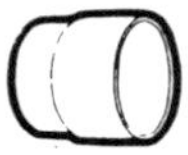

Fig. No.	Nominal Size
	3 x 2
	4
	4 x 1½
	4 x 2
	4 x 3
4801-RP & 5801-RP COUPLING REPAIR Plastic x Plastic	
	1½
	2
	3
	4
4801-2-F & 5801-2-F BUSHING Fitting x Plastic	
	1½ x 1¼
	2 x 1¼
	2 x 1½
	3 x 1½
	3 x 2
	4 x 2
	4 x 3
4801-2-7 & 5801-2-7 ADAPTER TRAP Fitting x Slipjoint	
	1¼
	1½
	1½ x 1¼
	2
4801-7 & 5801-7 ADAPTER TRAP Fitting x Slipjoint	
	1¼
	1½
	1½ x 1¼
	2
4803 & 5803 ADAPTER Plastic x Inside Thread	
	1¼
	1½
	2
	3
	4
5803-TPA ADAPTER TRAY PLUG Plastic x NPSM	
	1½

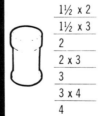

Fig. No.	Nominal Size
4803-2 & 5803-2 ADAPTER Fitting x Inside Thread	
	1¼
	1½
	2
	3
	4
4803-2-F & 5803-2-F ADAPTER Fitting x Inside Thread	
	1½ x 1¼
	2 x 1½
	3 x 1½
	3 x 2
	4 x 2
	4 x 3
5803-2-TPA ADAPTER — TRAY PLUG Fitting x NPSM	
	1½
4804 & 5804 ADAPTER Plastic x Outside Thread	
	1¼
	1½
	1½ x 1¼
	2
	3
	4
4804-2 & 5804-2 ADAPTER Fitting x Outside Thread	
	1¼
	1½
	1½ x 1¼
	2
	3
	4
4805 & 5805 ADAPTER SOIL PIPE Plastic x Spigot	
	1½ x 2
	1½ x 3
	2
	2 x 3
	3
	3 x 4
	4

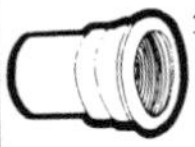

Fig. No.	Nominal Size
4805-N & 5805-N ADAPTER NO HUB SOIL PIPE Plastic x No Hub	
	1½ x 2
	2
	3
	4
4806 & 5806 ELBOW — 45° Plastic x Plastic	
	1¼
	1½
	2
	3
	4
4806-2 & 5806-2 ELBOW — 45° Fitting x Plastic	
	1¼
	1½
	2
	3
	4
4807 & 5807 ELBOW — 90° Plastic x Plastic	
	1¼
	1½
	2
	3
	4
4807-CL & 5807-CL CLOSET ELBOW — 90° Plastic x Plastic	
	4 x 3
4807-LT & 5807-LT ELBOW 90° LONG TURN Plastic x Plastic	
	1¼
	1½
	2
	3
	4
4807-V & 5807-V ELBOW 90° VENT Plastic x Plastic	
	1¼
	1½
	2
	3
	4

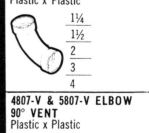

Courtesy Nibcoware Div., Nibco, Inc.

DWV Fittings for ABS and PVC drainage systems

Fig. No.	Nominal Size
4807-2 & 5807-2 ELBOW — 90° Fitting x Plastic	
	1¼
	1½
	2
	3
	4
4807-2-CL & 5807-2-CL ELBOW Fitting x Plastic	
	4 x 3
	4 x 3 W/Cap
4807-2-LT & 5807-2-LT ELBOW **90° LONG TURN** Fitting x Plastic	
	1½
	2
	3
	4
4807-2-V & 5807-2-V ELBOW **90° VENT** Fitting x Plastic	
	1½
	2
4807-7-LT & 5807-7-LT ELBOW **90° LONG TURN** Plastic x Slipjoint	
	1½
	1½ x 1¼
4807-9 & 5807-9 ELBOW — 90° **W/SIDE INLET** Plastic x Plastic x Plastic	
	3 x 3 x 1½
	3 x 3 x 2
4808 & 5808 ELBOW — 22½° Plastic x Plastic	
	1½
	2
	3
	4
4808-2 & 5808-2 ELBOW — 22½° Fitting x Plastic	
	3
	4

Fig. No.	Nominal Size
4810 & 5810 Y — 45° Plastic x Plastic x Plastic	
	1¼
	1½
	2
	2 x 1½ x 1½
	2 x 1½ x 2
	2 x 2 x 1½
	3
	3 x 3 x 1¼
	3 x 3 x 1½
	3 x 3 x 2
	4
	4 x 4 x 1½
	4 x 4 x 2
	4 x 4 x 3
4810-13 & 5810-13 Y — 45° Plastic x Female Thread x Plastic	
	1½
	2
	3
	4
	4 x 4 x 2
	4 x 4 x 3
4810-14 & 5810-14 Y — 45° Plastic x Plastic x Inside Thread	
	1½
	3
4810-17 & 5810-17 Y — 45° Inside Thread x Plastic x Plastic	
	3
	3 x 3 x 2
4811 & 5811 TEE Plastic x Plastic x Plastic	
	1¼
	1½
	1½ x 1¼ x 1¼
	1½ x 1¼ x 1½
	1½ x 1½ x 1¼
	2
	2 x 1½ x 1½
	2 x 1½ x 2
	2 x 2 x 1¼

Fig. No.	Nominal Size
	2 x 2 x 1½
	3
	3 x 3 x 1¼
	3 x 3 x 1½
	3 x 3 x 2
	4
	4 x 4 x 1½
	4 x 4 x 2
	4 x 4 x 3
4811-C & 5811-C TEE 2 WAY **CLEANOUT** Plastic x Plastic x Plastic	
	3
	4
4811-V & 5811-V TEE VENT Plastic x Plastic x Plastic	
	1¼
	1½
	2
	2 x 2 x 1½
	3
	3 x 3 x 1½
	3 x 3 x 2
	4
4812 & 5812 TY — LONG TURN Plastic x Plastic x Plastic	
	1¼
	1½ x 1½ x 1¼
4812-LR & 5812-LR TY LONG **RADIUS** Plastic x Plastic x Plastic	
	1½
	2
	2 x 1½ x 1½
	2 x 1½ x 2
	2 x 2 x 1½
	3
	3 x 3 x 1½
	3 x 3 x 2
	4
	4 x 4 x 2
	4 x 4 x 3

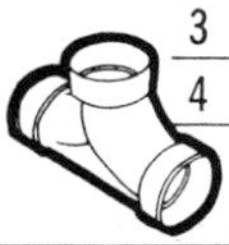

Fig. No.	Nominal Size
4812-13 & 5812-13 TY — LONG TURN Plastic x Inside Thread x Plastic	
	1¼
	1½
	3 x 3 x 1½
	4
	4 x 4 x 2
4812-13-LR & 5812-13-LR TY — LONG RADIUS Plastic x Inside Thread x Plastic	
	2
	2 x 2 x 1½
	3
	3 x 3 x 2
	4 x 4 x 3
4812-17 & 5812-17 TY — LONG TURN Inside Thread x Plastic x Plastic	
	3
	3 x 3 x 2
4814 & 5814 TEST TEE Plastic x Plastic x Cleanout W/Plug	
	1½
	2
	3
	4
4816 & 5816 CLEANOUT Fitting x Cleanout W/Threaded Plug	
	1¼
	1½
	2
	3
	4
4818 & 5818 PLUG Outside Thread	
	1¼
	1½
	2
	3
	4
4826 & 5826 PLUG Fitting	
	1½
	2

Courtesy Nibcoware Div., Nibco, Inc.

DWV Fittings for ABS and PVC drainage systems

Fig. No.	Nominal Size

4827 & 5827 CAP
Inside Thread

1¼
1½
2
3

5829 NIPPLE
Outside Thread x Outside Thread

3 x 3″ Length
3 x 4″ Length
3 x 5″ Length
3 x 6″ Length
3 x 7″ Length
3 x 8″ Length
3 x 9″ Length
3 x 10″ Length
3 x 12″ Length

4833-S & 5833-S UPTURN SINGLE STACK
Plastic x Plastic x Plastic

2
2 x 2 x 1½
3
3 x 3 x 2

4834 & 5834 Y — 45° DOUBLE
Plastic x Plastic x Plastic x Plastic

1½
2
2 x 2 x 1½ x 1½
3
3 x 3 x 1½ x 1½
3 x 3 x 2 x 2
4
4 x 4 x 2 x 2
4 x 4 x 3 x 3

4835 & 5835 TEE — DOUBLE
Plastic x Plastic x Plastic x Plastic

1½
2
2 x 2 x 1½ x 1½
3
3 x 3 x 1½ x 1½
3 x 3 x 2 x 1½
3 x 3 x 2 x 2
4
4 x 4 x 2 x 2
4 x 4 x 3 x 3

5835-A TEE — DOUBLE FIXTURE
Plastic x Plastic x Inside Thread x Inside Thread

1½
2 x 1½ x 1½ x 1½

4835-B & 5835-B TEE — DOUBLE FIXTURE
Inside Thread x Plastic x Plastic x Plastic

1½
2
2 x 1½ x 1½ x 1½
2 x 1½ x 2 x 2
3
3 x 2 x 3 x 3

4835-9 & 5835-9 TEE — DOUBLE W/ONE 90° SIDE INLET
Plastic x Plastic x Plastic x Plastic x Plastic

3 x 3 x 3 x 3 x 1½
3 x 3 x 3 x 3 x 2
4 x 4 x 4 x 4 x 2

4835-9-9 & 5835-9-9 TEE DOUBLE W/TWO 90° SIDE INLETS
Plastic x Plastic x Plastic x Plastic x Plastic

3 x 3 x 3 x 3 x 1½ x 1½
3 x 3 x 3 x 3 x 2 x 2
4 x 4 x 4 x 4 x 1½ x 1½
4 x 4 x 4 x 4 x 2 x 2

4836 & 5836 TY — DOUBLE LONG TURN
Plastic x Plastic x Plastic x Plastic

3
3 x 3 x 1½ x 1½
3 x 3 x 2 x 2
4

4837 & 5837 ELBOW — DOUBLE
Plastic x Plastic x Plastic

1½
2
2 x 1½ x 1½
3
4

4851 & 5851 CLOSET FLANGE
Plastic

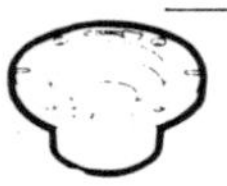

4
4 x 3

4851-A & 5851-A CLOSET FLANGE — W/INSERT
Plastic

4
4 x 3

4851-2-A & 5851-2-A CLOSET FLANGE — W/INSERT
Fitting

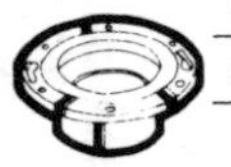

4
4 x 3

4851-3 & 5851-3 CLOSET FLANGE
Inside Thread

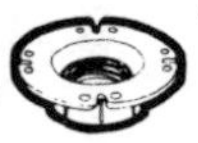

4 x 3

4851-4 & 5851-4 CLOSET FLANGE
Outside Thread

4 x 3

4860 & 5860 ELBOW — 60°
Plastic x Plastic

1½
2
3
4

4860-2 & 5860-2 ELBOW 60°
Fitting x Plastic

4

4861 & 5861 ELBOW 90° W/HIGH HEEL INLET
Plastic x Plastic x Plastic

3 x 3 x 1½
3 x 3 x 2

4861-LH & 5861-LH ELBOW — 90° W/LOW INLET
Plastic x Plastic x Plastic

3 x 3 x 1½
3 x 3 x 2

4870 & 5870 TEE — W/90° RIGHT & LEFT INLET
Plastic x Plastic x Plastic x Plastic x Plastic

3 x 3 x 3 x 1½ x 1½
3 x 3 x 3 x 2 x 2
4 x 4 x 4 x 1½ x 1½
4 x 4 x 4 x 2 x 2

4871 & 5871 TEE — W/90° LEFT INLET
Plastic x Plastic x Plastic x Plastic

3 x 3 x 2 x 1½
3 x 3 x 2 x 2
3 x 3 x 3 x 1½
3 x 3 x 3 x 2
4 x 4 x 4 x 1½
4 x 4 x 4 x 2

4872 & 5872 TEE — W/90° RIGHT INLET
Plastic x Plastic x Plastic x Plastic

3 x 3 x 2 x 1½
3 x 3 x 2 x 2
3 x 3 x 3 x 1½
3 x 3 x 3 x 2
4 x 4 x 4 x 1½
4 x 4 x 4 x 2

4876 & 5876 RETURN BEND TRAP W/CLEANOUT
Plastic x Slip Joint

1½
2

4877 & 5877 RETURN BEND TRAP
Plastic x Slip Joint

1½
2

4878 & 5878 RETURN BEND, W/CLEANOUT
Plastic x Plastic x Cleanout W/Plug

1½
2

Courtesy Nibcoware Div., Nibco, Inc.

DWV Fittings for ABS and PVC drainage systems

Fig. No.	Nominal Size

4879 & 5879 RETURN BEND
Plastic x Plastic

1½
2
3
4

4880 & 5880 P-TRAP W/CLEANOUT
Plastic x Slipjoint x Cleanout W/Plug

1½
2

4881 & 5881 P-TRAP
Plastic x Slipjoint

1½
2

4884 & 5884 P-TRAP W/CLEANOUT
Plastic x Plastic x Cleanout

1½
2

4885 & 5885 P-TRAP
Plastic x Plastic

1½
2
3
4

5885-TPA P-TRAP
Plastic x NPSM

1½

4891 & 5891 TRAP SWIVEL DRUM
Plastic x Plastic x Cleanout W/Plug

1½ x 3 x 6

4892 & 5892 P-TRAP W/UNION JOINT
Plastic x Slip Joint

1½
1½ x 1¼
(L. A Trap)

4892-3 & 5892-3 P-TRAP W/UNION JOINT
Inside Thread x Slipjoint

1½
1½ x 1¼

4895 & 5895 P-TRAP W/UNION JOINT
Plastic x Plastic

1½
2

5895-TPA P-TRAP W/UNION JOINT
Plastic x NPSM

1½

4895-3 & 5895-3 P-TRAP W/UNION JOINT
Inside Thread x Plastic

1½

828 ROOF FLASHING
Neoprene

1¼ x 1½
2
3
4

828-G ROOF FLASHING
Galvanized W/Neoprene Collar

1¼ - 1½
2
3
4

4898-2 SOLVENT — PVC
Clear With Applicator

¼ Pint
½ Pint
1 Pint
1 Quart

4899 PRIMER — PVC & CPVC
With Applicator

½ Pint
1 Pint
1 Quart

5898 SOLVENT — ABS
With Applicator

¼ Pint
½ Pint
1 Pint
1 Quart

5899 THINNER — ABS
Less Applicator

1 Pint
1 Quart

Courtesy Nibcoware Div., Nibco, Inc.

Index